THE ALICE B. TOKLAS COOK BOOK

THE ALICE B. TOKLAS COOK BOOK

ALICE B. TOKLAS

Foreword by

M. F. K. FISHER

Introduction by

RUTH REICHL

Illustrations by

SIR FRANCIS ROSE

HARPERPERENNIAL MODERNCLASSICS

NEW YORK • LONDON • TORONTO • SYDNEY • NEW DELHI • AUCKLAND

HARPER**PERENNIAL** ● MODERN**CLASSICS**

First Harper & Brothers hardcover edition published 1954; reissued Harper & Row, 1984.

HarperCollins books may be purchased for educational, business, or sales promotional use. For information, please email the Special Markets Department at SPsales@harpercollins.com.

FIRST HARPER PERENNIAL EDITION PUBLISHED 2010; REISSUED 2021.

Library of Congress Catalog Card Number 84-47602

ISBN 978-0-06-304380-0

21 22 23 24 25 LSC 10 9 8 7 6 5 4 3 2 1

CONTENTS

PUBLISHER'S NOTE TO THE 30TH ANNIVERSARY EDITION (1984)

"OH," SAID ALICE, "I COULDN'T DO THAT."

"Why not?" asked the young publisher from New York who had been trying to persuade her to write a book about her life with Gertrude Stein and about the many people and adventures they had shared.

"Because," said Alice in that cigarette-rough and sensuous voice, "Gertrude did my autobiography and it's done."

Since I could think of no response to that, I must have looked very sad; and since Alice had grown a bit fond of us in the years since Gertrude's death, she thought of something:

"What I could do," she said as tentatively as she was able, which was not very, "is a cook book." And then, "It would, of course, be full of memories." She kept her promise.

In the extraordinary book she then wrote, she says of the French, that, like their Bourbon kings: "They learn nothing, they forget nothing." Of Alice it can be said that she learned a great deal and forgot nothing. Into this modest (in size) volume, she put a generous sampling of what she had learned about food and cooking (French and American as she reinvented it), about the numerous and immensely varied and often very gifted friends and acquaintances whom she and Gertrude attracted, about life in Paris and in the country, about War and Occupation, about the U.S.A. in the 1950s, about servants

(French, Swiss, Finnish and Asiatic), and, of course, about the celebrated woman who was the love and center of Alice's life.

During the writing of the book there were several memorable moments for the lucky publisher, whose sole qualification for having anything to do with a book about food and cooking was a prodigious appetite for food (and wine) and a fascination for anything Alice did or said. One such moment came when Alice decided to test several of her recipes on the publisher and her old friend Thornton Wilder. For some reason, probably money, which she then had very little of despite the immensely valuable things on the walls, Alice decided to do without a maid, which meant the tiny and aging figure spent most of the lunch running back and forth between kitchen and dining room. Wilder, one of the world's politest men, stood up every time Alice did. When she had put a huge platter of fried chicken on the table and Thornton was still at half-mast, she tried to help him to some chicken, saying curtly: "Thornie, for God's sake, sit down and I'll give you some chicken. Which do you prefer; light or dark?"

The great man turned to his sister Isabel and asked: "Which is it I prefer?" "Oh, God," said Alice, "help yourself, dear Thornie." And back to the kitchen.

When the book appeared in 1954 it caused a small sensation, partly because of the fuss about hashish fudge, but mostly because of what it was: a unique book by a unique person. Which, of course, is why it's stayed in print for thirty years and why there's this new edition, about which I presume to repeat what Alice once wrote when inviting us to lunch two days after we had lunched: The tiny words said: "If this seems too soon to you, remember what the young man said: If perfection is good, more perfection is better."

Simon Michael Bessie

1954–1984 *17.vii.84*

FOREWORD

WHAT AN EXTRAORDINARY PERSON! MISS TOKLAS HAS BEEN AN INTEGRAL part of my life (sensate, thinking, sensuous, spiritual) since I was in adolescence. And when I was newly twenty-one and newly married, I could have met her.

My husband and I stopped in Paris in 1929 on our way to some three years in Dijon at the university, and we had an "introduction" to Miss Stein and Miss Toklas, but I could not bring myself to present it. I could not walk around the corner with the letter in my hand.

Many years and lives later . . . and the fact remains that never did I meet this strange person, except through what other people wrote and said about her.

A while ago I re-read *The Autobiography of Alice B. Toklas*, which the *Columbia Encyclopedia* says Gertrude Stein published in 1933 "as if by her secretary-companion." It is amazing, a literary tour-de-force, an almost great writer going with almost surgical sureness into the self of a loved one. It is amazing, because it feels and smells and is true; it is prescient. It is especially amazing for its detachment, its lack of vanity; it is not even condescending, the way a person is not condescending to his inner self. It is, for the lover who was Gertrude and who can be me or any chosen soul, the person who was/is Alice B. Toklas, and *no other.*

People who knew Alice or even met her casually—she often writes of "friends (whom we had never met) of friends of ours"—always knew her at once and forever, the way I did when I read her "autobiography," the way Gertrude did when, a few minutes after she saw Alice, she said flatly that they were married for life. People have told me that when this small ugly woman was in a room they were keenly aware of her, before they even recognized her as Miss Toklas. She seemed to send out waves of inaudible sound, like bells clanging somewhere in another space than ours. And since I first read her so-called life, I am like everyone else, and know almost more about her and with her than I am really entitled to.

This was as true when I last lived in Paris as it had been the first time, in 1929, when I often walked past her door and knew she was behind it. By 1967 I had lived long enough to shed some of my first timidity, but she was hospitalized, too remote ever to welcome me as she might once have done.

I should perhaps try to explain how it happened that I missed my one last chance to meet her. In about 1938, my second husband and I were living in a vineyard south of Geneva, and we knew that now and then we must escape from our cautiously Calvinistic life as foreign land-owners. We decided that we could afford, for a year or so, to rent two small rooms in the servants' quarters of the old Hôtel Continental in Paris, high in the attics above the Tuileries. We could leave books there, and perhaps some pictures, and city clothes. In the cold winters we could go up to La Capitale, escape the colder, duller, more structured life in Chexbres, feel warm and free before it was time to start spring planting and vineyard work. . . .

Death and war changed all that.

But when I was offered a summer job in France by Time-Life, more than twenty years later, I took it. I felt I must go back, and this was my chance to. I was scared about being in Paris alone, for the first time in my life. I'd been there countless times before, with parents, lovers, husbands, children. This time I was by myself, my *self*.

I asked to be lodged in the attic rooms my husband and I had planned to live in. By then they had become stylishly expensive, but thanks to the potent clout of my temporary employers I went straight from St. Helena, California, to the small, low-ceilinged cubbyhole we'd meant to use for our books, our workroom. Sounds arose filtered and thin through the one big window, and the thick green of the Tuileries in summer. And in the next room, where we had meant to sleep, lived a trespasser, a stranger who became my good true friend, an elderly writer named Janet Flanner.

I had for a long time admired her masterly letters from Paris in *The New Yorker*, and at first felt some of the shyness that had kept me so long from presenting my letter of introduction to the two ladies around the corner on the rue de Fleurus. And there Janet was, in our private special room! And indeed she was *there*, with her plain typing table, one beautiful cabinet of inlaid *boiserie*, always with a big fading garden bunch of roses or field flowers brought each weekend from her lover's country house, her little bathroom always hung with a drying elegant nightgown or some tiny high-style panties. There were perhaps a hundred books and no pictures, and her narrow bed made the little room seem almost austere, except for the hum of all Paris as it rose from far below, and the magnificent light that poured in and up from the Tuileries and the Seine and the Left Bank. (Not long after that summer, Janet had to move to another old hotel because the Continental was bought by a world airline, but she was never as truly "at home" as in the attic bedroom where once I had thought I would live forever.)

Janet was much spryer than I, but was used to deputizing her many disciples, so that I spent most of that summer happily puffing around Paris on errands for her, fending off her fans at concerts, sampling a new batch of Sancerre in a cool cellar under the Luxembourg, with an ancient vintner she had known for countless years. . . . It was fine, and instead of being alone and scared in Paris for the first time in my life, I was more alive and happy than I'd ever dreamed of, because of all the good people who had unwittingly prepared me for it. I felt strong with their strengths, so that the work I had to do for Time-Life, and all the wild errands and jobs Janet loaded on me, seemed child's play.

But there was one thing she refused flatly to let me help her with. She refused to let me join her. She refused to *include* me, even vicariously, when she went several times a week to the clinic in the suburbs where Alice B. Toklas lay like a sightless, speechless vegetable.

Janet said firmly and simply that I must not see her now. She said that Alice would refuse to let me come, if she could.

So I never met her. *Ever*, that is.

Almost every day I went with Janet to Fauchon or a couple of other fine pastry shops that made *napoléons* or truffles or *palmiers* that Miss Toklas had once approved of, and watched Janet pretend that she was not going to eat the little treats herself as she sat by the bed of her impotent old friend. Then we hurried to the Métro and I watched her run down the stairs on her stylish tiny feet, carrying her stylish tiny offering.

She had known Alice B. Toklas for decades, about as long as I had, but face to face, as I never would. She was filled with love, and with anger that her old friend must die poor and abandoned. I mourned for them both, and was glad that Janet would eat a dainty voluptuous tidbit as she sat faithfully in Miss Toklas' silent hospital room.

But how else would Alice have chosen to be, after Gertrude died in 1946? It is true that the years between then and her own withdrawal were full; some people said that she finally "came into her own" as a dynamic and important figure. This may have seemed true, whether or not there could have been some malice in it, and I know from several of her friends that she ran her life with spirit, and entertained well, and handled the increasing complications about Gertrude's papers and belongings with surprising skill, for a person who had spent most of her life as a willing shadow. Myself, I feel sure that she was simply proving that Gertrude Stein had taught her well about the art of survival, as one final and gracious proof of their shared confidence.

Of course I "know" exactly what Alice looked like, and so far have not seen a picture of her that matches my own inner ones. Most of them are timid about how ugly she was. She was probably one of the ugliest people anyone had ever seen, to draw or photograph. Her face was sallow, her nose was big or even huge, and hooked and at the same time almost fleshy, the kind that artists try not to draw. And she had a real moustache, not the kind that old women often grow, but the sturdy kind, which started when she was first going into adolescence. I don't think she ever tried to shave it, or have it plucked out or removed chemically or with hormones, as a woman might do today. She wore it unblinkingly, as far as I can tell, although of course as a person of unusual awareness she must have known that some people were taken aback by it. A friend of mine who admired her greatly, and often traveled with her in her last years, wrote that Miss Toklas wore her close-cropped hair, which stayed black well into her eighties, in bangs "faintly echoed by a dark down on her lip." This amuses me. It is typical of the general reaction to something that would have been unnoticed except for her obvious femaleness. Another friend said more aptly, or at least better for my own picture, that her strong black moustache made other faces look nude.

She had remarkable eyes, very large and lively, the kind that seem to send off sparks, that sometimes look glowing with an inner fire. Probably people who were intimidated at first by her fixed upon them with relief . . . that is, until they forgot their shyness in the deft, supple way she moved and talked.

She was a tiny person, not five feet tall, I think, and she dressed with a studied daintiness, except for the clunky sandals on her pretty feet. They were almost an affectation, and almost offensive, the kind that Raymond Duncan and his followers wore as they ambled along the sidewalks of the Left Bank, unnoticed except by awestruck tourists. They wore togas of hand-woven wool or cotton, depending on the weather, with thongs criss-crossed up their bare legs from their open sandals, and Isadora's brother always strode a few paces ahead, with a twinkling Cartier wristwatch on the arm that wielded his long shepherd's staff. Of course a few young students imitated these Chosen Few, at least in footwear, but it was always funny to see the sandals on Miss Toklas, below her fastidiously tailored suits, her fine silk blouses, even her loose, beautifully sewn house clothes.

She loved dramatic hats, and after Miss Stein's death she wore them oftener in rare gaddings . . . big extravagant creations with feathers and wide brims, and always the elegant suits and those clunky sandals. Nobody has ever written, though, that she looked eccentric. Perhaps it was because of her eyes. . . .

According to the *Autobiography*, she cooked and Gertrude wrote. And according to her own cookbook, written several years after Gertrude's death, she cooked what she remembered eating when she was a girl in San Francisco, because her friend was homesick for American dishes now and then, so that from the time they first settled into their apartment on the rue de Fleurus in 1910 they served their own kind of "soul food" every Sunday night. But Alice had never been a housewife, in our sense of the word, so that the ways she evolved her down-home dishes stemmed more from nostalgia than kitchen experience, and the longer she lived in France and hired cooks there, and ate there, the less recognizable her "corn-pone and apple pie" became.

Of course Stein, who seldom skipped a good meal and knew thoroughly the pleasures of a well-tended palate, never boiled water, much less an egg, as far as is known. Her "secretary-companion" tended to all that, and her delicious food kept Gertrude's ink flowing for all their long life together. It is even possible that Miss Toklas honestly *believed* that she only cooked while Gertrude worked, and at the end of her famous *Cook Book* she says that when she told two friends that she was going to publish a book, "The first one gaily responded, How very amusing. The other asked with no little alarm, But, Alice, have you ever tried to write. As if a cook-book had anything to do with writing."

And Alice B. Toklas honestly did not believe that it does, or even that it can. Probably she would shrug, and smile with pity and disbelief, to find that the cookbook she put together for publication in 1954 would immediately be recognized and then be issued again and again. It is of course a curiosity, for many reasons, like the fame of many of the artists and writers who came to talk and listen and eat in the Paris apartment of the two strange ladies from America. Alice knew why all of them were there, especially Gertrude. But she herself never believed that she could ever be more than their attentive loving nourishing shadow.

And yet probably not even Stein herself would have been able to write of a person she observed in a friend's house, "Like many first-rate women-cooks she had tired eyes and a wan smile." Miss Toklas made comments about the people and the recipes in her book exactly as she talked, so that she felt that her notes were not worthy of being called writing, which was to her a life apart, mostly occupied by Gertrude, with a few lesser geniuses lurking behind her mammoth shadow. And after Gertrude died, Alice went on talking, fortunately for all of us!

According to her, the *Cook Book* was written while she was laid up for several months with hepatitis. Her naturally sallow face turned pumpkin-yellow, and while she waited for her insides to stop heaving and churning at the thought of food, she satisfied her emptiness by remembering better days, as when the two ladies chugged around France in their ancient Model T Red Cross ambulance called Aunt Pauline. In 1916, Stein was "a responsible if not experienced driver. She knew how to do everything but go into reverse. She said she would be like the French Army, never have to do such a thing." And Aunt Pauline hauled food and wood and the wounded men dauntlessly, always forward. . . .

Later there was Godiva, so named because she came to them stripped of everything on her dashboard, naked as only a "two-seater open" Ford could be in France after World War I. And between her and the worn but willin' Aunt Pauline, the other ladies seemed to find good food at the end of every mission, wars or not. Tarte Chambord, saddle of mutton, peaches, grilled perch with fennel . . . Alice never really loved Godiva, but always admired her.

Once after the war the sturdy Ford carried the two ladies to Vence, where Alice spent long mornings in a friend's vegetable garden. "It takes a long time to gather enough very young green beans for eight to ten people," she found. Between the vegetables and the roses her mornings were full, and she said happily, "To me this pleasure is unequalled."

Godiva took them much further along in their good lives, and had what both the ladies felt was an infallible gastronomical nose for sniffing out fine country restaurants for them. She was at last retired, not as a revered, respected remnant of the First World War, as Aunt Pauline had been, but respected nonetheless.

Of course it takes more than a lengthy jaundice attack to make a decent cookbook, much less a minor masterpiece, and there is no doubt that plain loneliness after the death of Gertrude Stein, as Alice always refers to her, meant that time had to be filled, with the quiet rich dignity that the tiny old "secretary-companion" had always shown. So in 1954 she finished a delightful collection of memoirs, *Cook Book*, and then in 1958, to international astonishment, she agreed to publish another book about food, when she was well past eighty, *Aromas and Flavors of Past and Present.*

In those days editors were still gracious enough to confess to the sins and errors of their proofreaders, so that a list was often inserted in a nicely designed book, giving words that had been omitted and phrases that slipped past the copyreaders. The first edition of *The Alice B. Toklas Cook Book* has one of these quaint courtesies, listing ten ERRATA, and I feel quite sure that the author herself dictated it. Only three are really important. The rest correct misspellings or direct that "Gastronomique Guide" be changed to "Guide Gastronomique." It was easier and cheaper to be finicky then than in 1984!

And collectors will always want this curious book for one omission caused by our American Puritanism, as well as for its good printing and its miserably inept illustrations: it could not print Miss Toklas' recipe for Haschich Fudge for legal reasons.

By now, of course, the sticky candy that sounds a little like chopped fruit balls that children make for their relatives at Christmas has become more like cookies or brownies, but always named for

Miss Toklas and always made with marijuana. Her recipe, regretfully omitted in 1954 but reprinted in paperback in 1960, was contributed by her friend Brion Gysin, and calls for "a bunch of *canibus sativa*" pulverized. This plant is common in Europe, Asia, Africa, as hemp, and Miss Toklas kindly says that "in the Americas, while often discouraged, its cousin, called *canibus indica*, has been observed even in city window boxes."

I have never eaten one of our "Toklas fudge brownies," but am told they taste slightly bitter, depending on how much pot is put into them, and that (1) they are absolutely without effect and (2) they are potentially lethal. Her directions are more lyrical. She first says that "anyone could whip up [Haschich Fudge] on a rainy day," and continues, "This is the food of Paradise—of Baudelaire's Artificial Paradises: it might provide an entertaining refreshment for a Ladies' Bridge Club or a chapter meeting of the DAR. In Morocco it is thought to be good for warding off the common cold in damp winter weather and is, indeed, more effective if taken with large quantities of hot mint tea. Euphoria and brilliant storms of laughter; ecstatic reveries and extensions of one's personality on several simultaneous planes are to be complacently expected. Almost anything Saint Theresa did, you can do better. . . . "

Was this Alice B. Toklas talking, or Brion Gysin? The chapter called "Recipes from Friends," which gives their later, much-distorted recipes, holds several other deviations from her way of stating things. They simply don't *sound* Toklas-ian. Mary Oliver of London, for example, gives a terse recipe called "Birthday Ice Cream for Adults" that really sounds dreadful, and I like to believe that Miss Toklas included it because she was fond of Mary Oliver. (One of my grandmothers was named that too, but as a teetotalling Irishwoman she would never have considered adding a cup of rum to anything.)

During the German occupation of France in the Second World War, food was austere at best, with milk, butter and eggs almost unknown even in the country near Belley, where the two American ladies lived in precarious security. Meat was rationed: a quarter pound a week per person. At best, they lived in what Toklas called a "protracted, even a perpetual Lent." And it was then that she did what people have always done in times of hunger: she betook herself to "the passionate reading of elaborate recipes in very large cookbooks."

Often I have sent *Larousse Gastronomique* or an American kitchen bible like Mrs. Rombauer's *Joy of Cooking* to students working their way through college, or to men in prison, and they have nourished themselves in many ways from their printed rations. And I know now as well as I did thirty years ago that Alice B. Toklas' *Cook Book* would feed my soul abundantly if I could find no other nourishment, just as it would make me smile in the midst of sadness, and feel braver if I risked faltering. It is a good book, "abundantly satisfying, imagination being as lively as it is. . . ."

M. F. K. Fisher

Glen Ellen, California
5.iv.84

INTRODUCTION

ALICE B. TOKLAS WAS HUNGRY.

She sat shivering in the single room she could still afford to heat in the Paris apartment she had once shared with Gertrude Stein. Surrounded by Picassos, Matisses, and Cézannes, she contemplated her bare cupboard.

Food in postwar France was rationed. Black market prices were ruinous. And Alice B. Toklas was broke.

Stein had passed away six years earlier, leaving her companion of almost forty years in custody of her art collection. But the terms of the will specified that nothing could be sold without the permission of the trustees—and the trustees were dragging their feet. And so, at the age of seventy-five, Alice B. Toklas hatched a plan.

Americans were permitted access to the embassy commissary—where food, liquor and cigarettes were sold at bargain prices—provided they had a legitimate reason. "You may remember," she wrote to a friend, "I've been trying to maneuver to be admitted to the American commissary . . . it came to me if I could get recipes printed in some magazine I'd be as eligible as Richard Wright—so why not gather my recipes—make the cookbook and get a job."

It was the genesis of what went on to become one of the bestselling cookbooks of all time.

Toklas was a notable and noted cook, and she could easily have produced a straightforward book filled with a few fine recipes introduced by little headnotes. If all she wanted was access to food, she could have written a conventional cookbook, a traditional cookbook, one like all the others. But Alice B. Toklas chose to write a very different kind of book, and her reasons remain mysterious. What exactly was she up to?

Twenty years earlier Gertrude Stein, convinced of her own genius and frustrated by her lack of *"gloire,"* had set out to write a bestseller. The way to do that, she decided, was to write a book ordinary people could understand. To distinguish it from her difficult modernist writing—the real work—she wrote in Toklas' voice and called it an autobiography. She ended *The Autobiography of Alice B. Toklas* like this:

> About six weeks ago Gertrude Stein said, It does not look to me as if you were ever going to write that autobiography. You know what I am going to do. I am going to write it for you. I am going to write it as simply as Defoe did the autobiography of Robinson Crusoe. And she has and this is it.

Virtually everyone who ever met Toklas and Stein agreed that Stein filled the "autobiography" with stories that Toklas often told, capturing Toklas' voice so convincingly that many believed Toklas had actually written the book herself. "I am unable to believe that Gertrude with all her genius could have composed it," said painter Maurice Grosser, "and I remain convinced that the book is entirely Alice's work and published under Gertrude's name only because hers is the more famous." Nobody will ever know the truth about that, and it's probably not important. The lives of the two women were so closely intertwined that it's difficult to think of one without the other.

They came from similar backgrounds; both were raised in California by moneyed Jewish families, both were educated, well-traveled, and highly intelligent. But while Stein was a large, warm, outgoing and supremely confident woman, Toklas was tiny, birdlike and self-effacing. She was also rather stunningly ugly, with a huge beak of a nose and an unabashed black moustache. It was an ugliness with its own fascinating glamour, which Toklas played up by dressing in extraordinary clothing and extravagant hats. Her one beauty was her voice: poet James Merrill described it as "a viola at dusk."

They fell in love almost the moment that they met, and their passion never abated. "It was Gertrude Stein who held my complete attention, as she did for all the many years I knew her until her death, and all these empty ones since then" is how Toklas described her feelings. They never tried to hide their love, openly calling each other "lovey" and "baby"—and if they were ever troubled—by their families, their friends or the authorities—it is not mentioned in either of their writings. Stein called Toklas "wifey" or "baby precious" while Stein was "Mr. Cuddle-Wuddle." When Stein stayed up, writing late into the night, she left little notes beside her sleeping companion's pillow. She signed them "Y.D." or "Your Darling."

As well she might, for Toklas expended all her energy in caring for Stein. "It takes a lot of time to be a genius," Stein said as she allowed Toklas to do almost everything: cook, run the household, type the manuscripts, entertain the wives of the celebrated artists and writers who came to visit, and care for the many dogs that shared their lives. When a photographer wanted Stein to occupy herself in front of the camera, they had a hard time coming up with a single task she performed for herself. Everything he asked her to do—unpack a suitcase, answer the phone—turned out to be something Toklas did for her. In the end Stein said, "I like water I can drink a glass of water all right he said do that." In a memoir written just after Stein's death in 1946, their friend W. G. Rogers described Toklas this way: "She doesn't sit in a chair, she hides in it; she doesn't look at you, but up at you; she is always standing just half a step outside the circle. She gives the appearance, in short, not of a drudge, but of a poor relation, someone invited to the wedding but not to the wedding feast."

But appearances were deceiving: Alice B. Toklas was made of steel. Writer Naomi Barry remembered an afternoon in the kitchen with Toklas. "She once gave me some celery to string. After ten minutes I handed it back, satisfied that I had removed every fiber. She simply returned it saying tersely, 'I said string it.'"

Nobody knew the steely side of Toklas better than Stein, who sprinkled references to her partner's toughness throughout *The Autobiography of Alice B. Toklas*. Mentioning that St. Anthony, patron saint of lost objects, was a special favorite she noted, "Gertrude Stein's elder brother once said of me, if I were a general I would never lose a battle, I would only mislay it." At another point Stein, writing in Toklas' voice, says, "When I first met her in Florence, she confided to me that she could forgive but never forget. I added that as for myself I could forget but not forgive."

Listening to Toklas talk, I can't help wondering if the cookbook might have been a way for this unforgiving woman to get a bit of her own back. The careful reader will uncover many little hints. There is, for instance, the moment when she finds herself seated next to author James Branch Cabell. She is paralyzed with fear until Cabell leans over and puts her at ease. "Tell me Miss Stein's writing is a joke, isn't it?" he asked. "After that," Toklas says innocently, "we got on very well." And then, in her sphinx like fashion, she simply moves on.

There is also the matter of the book's ending.

Two friends are discussing the notion of Alice writing a cook-book. "The first one," Toklas notes, "gaily responded, How very amusing. The other asked with no little alarm, But, Alice, have you ever tried to write? As if a cook-book had anything to do with writing."

And then, bam, she's gone. The book is over, and you are left to ponder her meaning. Some people (among them M. F. K. Fisher) are convinced that this is Toklas being characteristically self-effacing. See, she's telling you, don't take this book too seriously. Ignore the words and stick to the recipes.

Really?

Toklas was an ardent, lifelong reader of cookbooks. She liked them so much that Stein's annual Christmas present was always a cookbook. Even at the height of the war, with great effort and at enormous expense, Stein managed to procure a rare cookbook for her partner. They read it together. Indeed, as Toklas sat down to begin her own book, she wrote to a friend, "Thank you for letting me see your mother's cookbooks. . . . I enjoyed them immensely—the one I liked best naturally had the most extravagant recipes—nothing one could possibly afford but that made reading it more romantic and more of an adventure. It has given me an idea for my own humble effort. A cookbook to be read. What about it."

This is indeed a cookbook to be read. Open to a page—any page—and I promise you will find at least one sentence that opens your eyes and makes you see the world in new ways. Has there ever been another writer who would tell you that a custard has "the colour of its flavour" or that a soup will "come beautifully limpid"? Her voice is remarkable. Alice B. Toklas dreams food onto the page with the words of a poet. You can never anticipate what she is going to say—she surprises you at every turn; it is one of the great pleasures of this book.

When I was growing up my mother owned only two cookbooks. Sadly, the one that lived in the kitchen, the one she used, was Poppy Cannon's *The Can-Opener Cook Book* (which, like this one, was published in the early fifties). Mom loved that book, loved Ms. Cannon's concoctions; she especially cherished a cheese soufflé made with Velveeta and canned white sauce. I doubt she ever opened Alice's book, but had she done so she would have been appalled by the arcane ingredients: truffles, hare, nasturtium leaves! Even Alice's simplest recipes would have boggled Mom's mind: her rice pudding calls for fourteen egg yolks. No wonder Alice's book remained, its pages pristine, among the tomes on the living room bookshelves.

Why did my parents even have a copy? I never thought to ask, but it sat surrounded by the many Gertrude Stein books my father had designed, and I imagine he purchased it when he was working on *What Is Remembered* by Alice B. Toklas. All I know is that I was a teenager when I stumbled upon it one rainy afternoon. By then I'd started cooking, and I opened the book, eager for new recipes. Instead, I found myself captivated by a chapter called "Murder in the Kitchen." "The carp was dead, killed, assassinated, murdered in the first, second and third degree," I read. I was hooked.

I turned the pages, entranced by the tales of feeding Picasso and

Picabia, and by Toklas' descriptions of a French countryside where persimmon trees with orange fruit are silhouetted against a brilliant sky. But then I came to this sentence and simply stopped, staring down at the page, wondering what it meant. Toklas is in Seville, searching for an authentic recipe for gazpacho. She enters a bookstore. "Cook-books without number, exactly eleven, were offered . . . " she says. I read the words again. Without number is not exactly eleven. Especially to a recipe writer. I felt as if Toklas was whispering in my ear, sending me a secret message. Pay attention, she was telling me. Everything here is not exactly as it seems.

The food she described sounded delicious, but to be honest, the recipes worried me. Dad's first edition had a long list of errata glued to the first page. How could I trust a book that began by pointing out that a recipe for Green Peas à la Française was lacking four cups of shelled peas? Or that the scampi recipe neglected to mention that you had to cook the scampi? The one recipe I did attempt was a total disaster: it was for some little cakes called *visitandines*, and the editors had not noticed that the sugar was missing. (Unlike the omissions in the errata, that mistake has never been corrected.)

It was my loss. Years later I learned that many of my favorite cooks revered Alice Toklas. Richard Olney was obsessed with her, as was Alice Waters; indeed, one of the most famous meals at Chez Panisse was Jeremiah Tower's tribute to Toklas. I wish I'd been there: he served mushroom sandwiches, sole mousse with virgin sauce, gigot de la clinique, wild rice salad, and a tender tart. James Beard felt that the secret of her talent was "great pains and a remarkable palate," and I found that was true. As I began to cook the recipes I began to slow down, pay attention and take great pains. More than that, I began to trust my own palate.

But although I may have been late to the recipes, Toklas had given me something even more valuable: an entirely new way to think about cooking. She was writing about so much more than how to make a decent dinner. This book made me want to invite interesting people over to eat, find new markets to explore, plant a garden, travel the world. This was a cookbook that made me hungry for life.

I thought about all that when, at twenty-two, I sat down to write a cookbook of my own. I was dirt-poor, living in a loft on the Lower East Side of New York City, and totally untrained, but I somehow managed to persuade a publisher to let me write a cookbook. I knew from the start that it had to be more than mere recipes; I envisioned a book that would make readers want to go out and explore the world around them, tasting as they went. Looking back I see that what I produced was a young and very naïve version of Toklas' book. I was not consciously aware of it, but I wrote about the joy of going to the market, the pleasure of following the seasons, the happiness that comes from cooking for friends. Like Toklas, I wrote a book that is very personal and very much a product of its time. Had I never read Toklas I probably would not have had the courage to allow my book to go meandering off in strange directions, but thanks to her I included chapters on anything I fancied: one on lemons, one on birthdays, even a chapter of recipes contributed by friends. And like Toklas, I wrote about my artist friends: at the time I was teaching one of Warhol's "superstars" to cook, and that too made its way into the book.

But Toklas' influence on me did not end with the format of the book. Because while most people come to Alice's book to read about Matisse and Hemingway, and to fantasize about a long-vanished Paris, my own response was quite different. I enjoyed seeing a long-gone France through Toklas' eyes, liked visiting France before, between and just after the wars. But the chapter that truly captured my imagination was the one about the two women eating their way across America.

After living in France for more than twenty-five years, Stein and Toklas came home. It was 1934, and *The Autobiography of Alice B. Toklas* had been such a huge success that they embarked on a triumphant cross-country tour. Stein was dreading it. *What*, she wondered, *would they eat?* They heard terrible tales of tinned vegetable cocktails and canned fruit salad. They almost didn't come.

Things do not begin well. Although their first meal in New York is "very good in its way," Stein decides to limit herself to a strict pre-lecture regime of oysters and honeydew melon. Toklas does not approve. "From the beginning," she sniffs, "the ubiquitous honey-dew melon bored me."

But as they travel west, she is slowly seduced by the food. There is a gorgeous turtle soup that begins with sun-dried turtle meat. Lamb is basted with sprigs of mint as it turns on a spit. The market in New Orleans is so wonderful she "would have to live in the dream of it for the rest of my life." There she also encounters Oysters Rockefeller, which she says "makes more friends for the United States than anything I know." In Texas she encounters the most beautiful kitchen she has ever seen, and by the time they reach California to indulge in "gastronomic orgies" (crabs, avocados, rainbow trout in aspic) she is calling it "God's own country."

This America was even more exotic and mysterious to me than the France described in the book. It was certainly not the America I knew. Like most children of my generation I had come to think of our national fare as little more than one giant hamburger stand. Now I began to wonder if all those delicious foods were still out there, waiting to be discovered. What Toklas had inspired in me was a desire to get to know my own country. Eager to find Alice's America, I hit the road in an attempt to unearth regional fare. It was the early seventies, and I was tired of being told that the only food worth eating came from somewhere else. Curiously, this woman who had chosen to spend her life in France made me want to know more about American food. My own cookbook became a kind of ode to the America of its time, and I will always be grateful that Toklas sent me on that journey. It turned out that there was, indeed, an American cuisine, and I was eager to discover it before it vanished forever.

Toklas and Stein spent seven months in America, growing so enamored of their native country that Toklas wrote, rather wistfully, that nothing might ever equal this adventure. Thornton Wilder offered to find them a new home in Greenwich Village and they briefly considered staying. I wish they had. For they were not just two aging American women. They were Jewish. They were openly lesbian. It was 1937. And Hitler was on the rise.

Later, they were urged to leave by their families and local officials. But they stayed on. In an article in *The Atlantic Monthly* in 1940, Stein explained why. "We telephoned to the American consul in Lyon and he said, 'I'll fix up your passports. Do not hesitate—leave.' But the next day . . . I said to Alice Toklas, 'Well, I don't know. Moving back would be awfully uncomfortable and I am fussy about my food. Let's not leave.'"

And as Toklas tells it, the Occupation wasn't so bad. They left Paris for the Bugey (near the Swiss border), and life went on. They gave their meat rations to the dogs and ate crayfish. At birthday parties they indulged in trout, braised pigeons and baron of lamb. They bought food on the black market. The recipes roll off her pen. It is easy to believe that Stein was telling the truth when she told Eric Sevareid the war years were "the happiest years of her life." Long afterward Toklas herself said wistfully, "Though Hitler and the presence of the Occupants was a menacing nightmare, I was happier then than today."

These are remarkable admissions from two Jews in the time of Hitler. How did they survive? You will look in vain for any clues in this book. Toklas talks of getting food from friends in the Resistance and the occasional soldiers—German, Italian—were foisted upon them. But for the most part, life went on.

How is this possible? The answer isn't pretty. In 1934 Gertrude Stein suggested that Hitler should be given the Nobel Peace Prize because he was getting rid of Jews and leftist dissidents and making Germany a more peaceful place. Apologists have tried to pass this off as ironic, but there's no explaining away the work Stein did for the head of the Vichy regime, Marshal Pétain. Stein translated Pétain's speeches into English with the hopes of getting them published (they never were). She even described herself as a propagandist for Pétain and continued to praise him after the war.

Did she do this to save their skins? We don't know. What we do know is that she was protected by Bernard Faÿ, a close friend and Vichy official who had translated *The Autobiography of Alice B. Toklas* into French. In his own memoir Faÿ wrote about talking of Stein to Pétain. "Before the meeting ended the Maréchal dictated a letter to the sous-prefect at Belley, entrusting Gertrude Stein and Alice Toklas to his care, and directing him to see to it that they had everything needed to keep warm during the winter, as well as ration coupons for meat and butter. I came to Vichy quite regularly and I telephoned the sous-prefect to remind him of his instructions. During this horrible period of occupation, misery and nascent civil war, my two friends lived a peaceful life."

After the war Faÿ was sentenced to life in prison for the things he had done during the war.

You probably wish I hadn't brought this up. Frankly, I wish I hadn't either.

But the truth is that our views of Gertrude Stein have been altered by our modern understanding of her collaboration with the Vichy regime. We read her differently than we once did. However, none of the vast literature about Stein's unpleasant politics says anything about Toklas' complicity. As in so much else, she stays in Stein's shadow. Alice has been given a pass; after all, she only wrote a cookbook.

But we can't have it both ways. We can let Alice remain a minor character—the little woman who didn't think for herself, whose politics don't matter—and allow her book to remain in the shadows too. But if we agree that her words matter—that a cookbook has something to do with writing—then we have to take a clear-eyed view of her legacy.

That casts a cloud over this book. My apologies. But history has not been kind to Alice B. Toklas. She is always the ugly stepsister, the afterthought, the one whose voice was stolen. The last years of her life were long (she died at the age of eighty-nine), and lonely, and she lost almost everything, including the beloved Paris apartment. But the one thing she would not give up was her guardianship of Stein's legacy.

Nobody has done the same for her. Now, pondering her words, I can't help wondering if she would prefer to live forever in the shadows, someone dimly seen, who did nothing more than produce a pleasant little cookbook in what *Time* magazine dismissed as her "prattle." Or would she rather go down in history—in all her fierceness, complexity and complicity—as the woman who changed the way we think of cookbooks?

Alice herself may have answered that question. In her final book she wrote about the last time she saw Gertrude Stein. The two are at the hospital. "I sat next to her and she said to me early in the afternoon, What is the answer? I was silent. In that case, she said, what is the question?"

Ruth Reichl

New York, NY

2020

A WORD WITH THE COOK

AS COOK TO COOK I MUST CONFIDE THAT THIS BOOK WITH ITS MINGLING OF recipe and reminiscence was put together during the first three months of an attack of pernicious jaundice. Partly, I suppose, it was written as an escape from the narrow diet and monotony of illness, and I daresay nostalgia for old days and old ways and for remembered health and enjoyment lent special lustre to dishes and menus barred from an invalid table, but hovering dream-like in invalid memory.

Illness sets the mind free sometimes to roam and surmise. Though born in America, I have lived so long in France that both countries seem to be mine, and knowing, loving both, I took to pondering on the differences in eating habits and general attitude to food and the kitchen in the United States and here. I fell to considering how every nation, for the matter of that, has its idiosyncrasies in food and drink conditioned by climate, soil and temperament. And I thought about wars and conquests and how invading or occupying troops carry their habits with them and so in time perhaps modify the national kitchen or table.

Such speculations led me to rout about among my huge collection of recipes and compile this cook-book. I wrote it for America, but it will be pleasant if the ideas in it, besides surviving the Atlantic, manage to cross the Channel and find acceptance in British kitchens too.

Paris, 1954 A.B.T.

THE ALICE B. TOKLAS COOK BOOK

I.

THE FRENCH TRADITION

THE FRENCH APPROACH TO FOOD IS CHARACTERISTIC; THEY BRING TO their consideration of the table the same appreciation, respect, intelligence and lively interest that they have for the other arts, for painting, for literature and for the theatre. By French I mean French men as well as French women, for the men in France play a very active part in everything that pertains to the kitchen. I have heard working men in Paris discuss the way their wives prepare a beef stew as it is cooked in Burgundy or the way a cabbage is cooked with salt pork and browned in the oven. A woman in the country can be known for kilometres about for the manner in which she prepares those sublimated dumplings known as *Quenelles*, and a very complicated dish they are. Conversation even in a literary or political *salon* can turn to the subject of menus, food or wine.

The French like to say that their food stems from their culture and that it has developed over the centuries. It has its universal reputation for these reasons and on account of the mild climate and fertile soil.

We foreigners living in France respect and appreciate this point of view but deplore their too strict observance of a tradition which will not admit the slightest deviation in a seasoning or the suppression of a single ingredient. For example, a dish as simple as a potato salad must be served surrounded by chicory. To serve it with any other green is inconceivable. Still, this strict conservative attitude over the years has resulted in a number of essential principles that have made the renown of the French kitchen.

French markets without deep freezing are limited to seasonal produce which is however of excellent quality with the exception of beef, milk and a few fruits. Even the common root vegetables, carrots, turnips, parsnips and leeks (the asparagus of the poor), are tender and savoury, olive oil and butter are abundant and of a high grade and bread is nourishing and delicious.

Wars change the way of life, habits, markets and so eventually cooking. For five years and more the French were deprived of most of their foodstuffs and were obliged to use inferior substitutes when they could be found. After the Liberation the markets very slowly were supplied with a limited amount of material. The population had been hungry too long, they had lost their old disciplined appreciation of food and had forgotten or were ignoring their former critical judgment. So that even now French food has not yet returned to its old standard.

The crowded continent of Europe on which wars are fought inevitably suffers more privations than we do. Restrictions aroused our American ingenuity, we found combinations and replacements which pointed in new directions and created a fresh and absorbing interest in everything pertaining to the kitchen.

The French are indifferent to these new discoveries of ours, to the exact science that American cooking has become, to our time- and labour-saving devices. Nor do they like the food that issues from our modern kitchens. They say that it is either too imaginative or too exotic. One may say of the French what was said of their Bourbon kings: they learn nothing, they forget nothing. Since the war we Americans have learned a great deal from various sources and as teaching is natural to us we would like to share our knowledge.

French cooking for the use of American or British women is not hedged around with as many difficulties as most of them suppose. If they permit themselves to indulge in national prejudices they should admit the same privilege for the French. For example the French use butter, and of an excellent quality, for nearly all their cooking, not only because it gives a flavour that no substitute does but because it "marries" as they say, that is it amalgamates all the flavours of the dish to be prepared as well as thickening the sauce. Which brings us at once to a fundamental difference between

French and American cooking. The famous five basic sauces do not prevent French cooking from being dry whilst American cooking is moist though devoid of sauces. The French drink wine with their lunch as well as with their dinner. Americans drink little wine if any with their meals but there are at least a dozen beverages from which they can choose an accompaniment to their food if they want to. Four of the five basic French sauces are certainly unknown even by name to half the population of France. Almost any Frenchwoman can prepare a white sauce, frequently and erroneously called by Americans a cream sauce. The French being realists look facts in the face and only call a sauce a cream sauce when it is made with cream. Some French sauces have a small amount of cream in them but that does not make them a cream sauce. The French never add Tabasco, ketchup or Worcestershire sauce, nor do they eat any of the innumerable kinds of pickles, nor do they accompany a meat course with radishes, olives or salted nuts. Respect for the inherent quality and flavour of each ingredient is typical of the French attitude to food and it gives a delicacy and poignancy to their cooking. Their discreet use of herbs is to be remarked. This restraint, *le juste milieu* they call it, the golden mean, is what makes them not only good cooks but good critics of food.

French cooking is founded upon the discoveries made in the seventeenth century when suddenly everyone who could afford it became interested in food as a fine art. It was a century of advancement in the art of living and the art of cooking was greatly refined and widened. Expressing its time as any original endeavour must do, French cooking underwent the influences of the lavish eighteenth century and the extravagances of the nineteenth century. The first half of the twentieth century has been too disturbed by two major wars to have yet declared itself.

To cook as the French do one must respect the quality and flavour of the ingredients. Exaggeration is not admissible. Flavours are not all amalgamative. These qualities are not purchasable but may be cultivated. The *haute cuisine* has arrived at the enviable state of reacting instinctively to these known principles.

What is sauce for the goose may be sauce for the gander but is not necessarily sauce for the chicken, the duck, the turkey or the guinea hen.

BOEUF BOURGUIGNON (1)

(Beef Stew as cooked in Burgundy)

2 lbs. of shoulder of beef without bones cut in squares of about 2 1/2 inches.

3 tablespoons lard.

3 1/2 ozs. salt pork cut in small squares.

12 small onions.

1 tablespoon flour.

2 cups old dry Burgundy red wine.

1 clove garlic, a bouquet tied together of about 2 inches orange peel, a bay leaf, a small sprig of thyme, a sliver of nutmeg. Salt, no pepper.

Melt the lard in a Dutch oven; when it smokes brown the salt pork and remove, brown the onions and remove. Then place the pieces of meat side by side and brown on all sides. Add the salt pork and flour, stirring with a wooden spoon. Add the wine well heated, stir well. Add the clove of garlic, the orange peel and the bouquet. Cover hermetically and cook over low flame for 3 1/2 hours. If the wine and juice have evaporated add a very small quantity of boiling water at a time. Add onions and cook for 15 minutes longer. Remove bouquet. Place meat and gravy on serving dish, sprinkle chopped parsley over top.*

Another version of this admirable dish will be found in Chapter IX.

* *Note.* A Dutch oven is a deep, round pot with a well-fitting cover and is used in the oven or over a low flame or hot-plate.

QUENELLES

(a short cut)

1 1/2 cups concentrated chicken broth.
1/2 cup butter.
1 1/2 cups sifted flour.
Salt, pepper, a pinch of nutmeg.
Yolks of 5 eggs.

Bring the concentrated chicken broth and butter to a boil in a saucepan. As soon as it comes to a boil remove from the heat and at once put into the saucepan as quickly as possible the sifted flour. With a wooden spoon working rapidly stir until it is perfectly smooth. Then place on a very low flame continuing to stir vigorously until the paste leaves the sides and bottom of the pan clean and small beads appear upon the surface. Remove from heat and cool for about 10 minutes. Then add the yolks of eggs, one at a time, beating each one with a high stroke that allows as much air as possible to enter. This should take about 20 minutes. Cover and put aside in a cool place, but not in the refrigerator, for 2 or 3 hours. Half an hour before time to serve put a large saucepan of water to boil. On a well-floured table take small pieces of dough and roll into finger lengths and gently drop into the boiling water. Reduce the heat—they should poach not boil. When they rise to the surface and turn over remove pan from the flame and cover. This will cause them to swell. A few minutes will suffice for this. Remove gently with a flat perforated spoon. Serve in a cream sauce or in a mushroom sauce or a combination of both or surrounding fricasseed chicken or with a veal roast.

SUGGESTIONS LEARNED FROM THE FRENCH

About the use of wine in cooking: Add red wine to beef, white wine to chicken, veal and pork. White wine to sauces made with cream.

Two tablespoons of cognac lighted and added to beef and mutton give an indefinable flavour. Mutton roasted and basted with port is out of this world. Try it.

Cream should be added to sauces at the last moment, only in time to heat thoroughly—it should not boil. It is tilted in the saucepan, not stirred.

A piece of butter added to a sauce at the last moment, also tilted and not boiled, makes the sauce unctuous and lightly thickens it.

II.

FOOD IN FRENCH HOMES

FOOD DIFFERS MORE FROM DAY TO DAY IN FRANCE THAN IT DOES IN THE United States, not only in the variety of each day's menu but in the choice of the menu according to the persons to whom it is to be served. In the United States there is less difference between the choice for the family's menu and one for guests than there is in France. For example, this is the menu at a lunch party to which we were invited at a house whose mistress was a well-known French hostess and whose food was famous.

Aspic de Foie Gras
Salmon Sauce Hollandaise
Hare à la Royale
Hearts of Artichokes à la Isman Bavaldy
Pheasants Roasted with Truffles
Lobster à la Française
Singapore Ice Cream
Cheese
Berries and Fruit

This copious lunch was accompanied by appropriate and rare wines. There were at table ten guests and six of the family. The fine linen and beautiful crystal, porcelains and silver were of the same quality as the menu.

HEARTS OF ARTICHOKES À LA ISMAN BAVALDY

Prepare 12 artichokes by cutting the leaves to within 2 inches of the heart. As each one is cut, put it into a recipient of cold water to which the juice of 1 lemon has been added. When all the artichokes are ready, shake them well to clean them in a quantity of running water. Put them at once in a large saucepan of furiously boiling water to which 1 teaspoon salt and 1/2 teaspoon cardomom seeds and the juice of 2 lemons have been added, and cover. There should be enough water to float the artichokes until they are tender, about 25 minutes according to size. As soon as a leaf can be removed easily, remove from flame, drain at once, and put into recipient of cold water and under running water. When water is tepid remove artichokes, drain, gently remove all leaves and the chokes. Trim around the hearts if necessary. The leaves can be scraped with a silver spoon and mixed with a little cream to be used in an omelette or under mirrored eggs. Boil 3 lbs. of small green asparagus tied in bundles in a covered saucepan of salted water. Cover and boil for about 15 minutes or until tender, but be careful not to overcook.

After soaking a sweetbread weighing about 1 lb. in cold water for 1 hour, boil in water to which 1/2 teaspoon salt, 2 shallots and 6 coriander seeds have been added. Boil covered for 20 minutes. Plunge in cold water and when cool enough, remove tubes and skin. Strain with potato masher through strainer. Put 2 tablespoons butter in frying pan, when the butter begins to bubble reduce flame. Put the sweetbreads in frying pan. Stir constantly until they are well mixed with butter. Sprinkle on them 2 tablespoons flour. Mix thoroughly. Then slowly add 2 cups dry champagne. Cook gently until this sauce becomes stiff.

Cut the asparagus within 2 inches of the tip. With the left hand, hold an asparagus upright in the heart of an artichoke while a wall of the sauce is built around it with the right hand. The tips of the asparagus should show about 1/2 inch above the sauce. Cover the sauce with a thick coat of browned breadcrumbs. Pour 1 tablespoon butter over each asparagus tip and the breadcrumbs. Place the artichokes in a well-buttered fireproof dish and brown in preheated 425° oven for 1/4 hour.

It does not take as long as it sounds to prepare this dish. The lemon, champagne and coriander seeds give an ineffable flavour.

SINGAPORE ICE CREAM

Stir 1 1/2 cups sugar with 12 yolks of eggs until they are thick and pale yellow. Slowly add 4 cups hot cream in which a vanilla pod cut in half vertically has been steeping. Mix thoroughly, pour into saucepan and stir constantly with a wooden spoon over lowest heat until the spoon is thickly covered. Remove from heat and pour through a fine sieve into bowl. Remove vanilla and wash the two pieces well in cold water. They may be used again. If you do not use vanilla bean, add 1 tablespoon vanilla extract. Stir the mixture from time to time until cold. Before putting to freeze, stir in 1 cup diced ginger as completely drained as possible of its syrup and 1 cup not too finely chopped blanched pistachio nuts. Then mix in 2 cups whipped cream. Put to freeze. It is not necessary to stir during freezing. When frozen take out of mould and decorate with 1 cup whipped cream flavoured with 2 tablespoons ginger syrup.

Sixteen of us sat down to table, and from these two dishes it is not difficult to appreciate the effort and expense of the menu prepared for guests. Nor is it difficult to appreciate the surprise with which we received this menu, when we were asked informally to stay for supper one evening.

Lentil Soup
Mirrored Eggs
Cold Ham with Lettuce Salad
Purée of Spinach with Croûtons
Cheese
Berries and Fruit

The table was elaborately decorated with hothouse flowers, the linen, crystal, porcelain and silver were beautiful, though not as precious as for the lunch party, and the service was impeccable. The plates were changed for each of the six courses, the knives and forks were not except once for cheese and again for fruit. The knives and forks between courses were placed on silver rests. Two of the members of the family—we were the only guests—on either side of me methodically between courses wiped their knives and forks on pieces of bread. The contrast of the two menus was a revelation of the way life was led in a French family of fashion.

There is for the French no way of understanding the American habit of having such attractively furnished and arranged kitchens as not only to make it possible but pleasant to eat one's meals in them. To them a kitchen is a room in which a great deal of preparation for cooking, as well as the cooking, takes place. The walls of the kitchen are therefore covered with suspended pots and pans and kitchen utensils, and the tables have at least two mortars and pestles and endless bowls, graters and sieves in evidence. And all this without any disorder but lacking the taste that prevails in the other rooms of the home. Taste there undoubtedly is—the cooking-culinary-gastronomic taste. It is a puzzlement to Americans.

An old friend of ours, living alone for some years now in conditions requiring the strictest economy, has finally accepted a modern conception of working in her home. It has become a pleasure and no longer a drudgery. For a long time she brought the food that she cooked from the kitchen to a well-set table in the dining-room, course by course. She now eats not only in the kitchen but in what she calls the Anglo-Saxon manner. The meat course is served with the vegetables and potatoes or salad. She moved a fine seventeenth-century cupboard into the kitchen, in which she keeps all that is necessary to serve a meal for four people—she does not have more than four guests as many times as that a year. In her kitchen, where she now reads, sews and writes letters in either of two fine winged arm-chairs, I have eaten a

BRAISED GROUSE

Clean the grouse, empty the cavity and wash it with a little milk. Pour the milk out. Put the grouse in a bowl and pour over it 1 cup milk. Leave it in the milk for at least an hour, turning it from time to time. Then dry and skewer the grouse so that the wings and legs remain neatly in place when served. Remove the skin if there is any on a slice of back fat of pork large enough to cover the grouse. Tie the lard securely around the grouse. Melt 1 tablespoon butter in a heavy saucepan or iron pot. Brown the grouse uncovered on all sides over medium flame for 1/4 hour. Skim off some of the fat, add salt and pepper, 1 cup cream and simmer covered for 20 minutes. Then add the juice of 1/2 lemon. After this do not allow to boil. Place grouse on serving dish and pour over it the sauce. It is a delicious way to prepare grouse, and had lost none of its savour because my friend had shot it. She goes off intrepidly on her bicycle for miles to shoot a bird or catch a fish. These economies and pleasures permit her to keep a small but excellent wine cellar.

She made an original but luscious *purée* to eat with the grouse.

PURÉE OF CELERY ROOT (CELERIAC) AND POTATOES

Wash and peel a celery root weighing 1 lb., remove all fibres, cut in large cubes. Boil in salted water until the tines of a fork enter into it easily. Wash 3/4 lb. unpeeled potatoes and steam. Use the same test for them as for the celery root. Peel and mash 4 potatoes with the celery root and 1 hard-boiled egg, put through a strainer with a potato masher. Add 3 tablespoons butter,

1/2 teaspoon salt, a pinch of pepper, and heat over asbestos mat until hot, stirring from the bottom so that it does not burn.

This *purée* makes an equally novel salad. Instead of heating in butter add 1/3 cup cream, place in a mound on a flat plate, cover with a thick mayonnaise in which 1 teaspoon lemon juice has been mixed and surround the mound with hearts of lettuce.*

In France it is not unusual for some man in the family not only to be interested in the menu and the cooking of it but occasionally to wish to supervise or even cook a dish. This raises the standard of cooking in the home, the mistress is spurred to greater effort by a constant gentle criticism. Women are not supposed in France to be *gourmets.* It is rare to find a woman whose taste in wines is as keen and subtle as a man's. There are of course exceptional cases, like that of a daughter who, instead of a son, inherited her father's famous wine cellar. At her table the *maître d'hôtel* presents each dish for this lady's inspection before serving it to the guests. At a large lunch party to my surprise after carefully examining the aspic of foie gras, she, with a brusque gesture, brushed it aside. What it was that at a glance did not satisfy her is a question that still torments me. She has needed no man's criticism to keep her culinary taste up to an almost solitary peak.

One of my friends, an admirable housekeeper and provider, is not so attentive to the kitchen and not so sensitive to the preparation of food. It is her husband who is a *gourmet* and a student of gastronomy. He is constantly delving into books for more information concerning how, when and why cooking developed as it did and became what it is. This in a roundabout way has influenced the cooking his wife superintends. A meal in their home may include a recipe from *The Treasure of Health* printed in 1607 with a preface by Jean-Antoine Huegetan.

* *Note.* The variety of celery grown exclusively for the edible properties of its root is known as celeriac.

The recipe is no other than our present *filet de sole à la meunière*, though it is differently worded.

THE TROUT

Small trout emptied and the scales removed, cut from head to tail on either side and from the bone. Rub some salt in them and put aside for 2 hours. Then rub excess salt from the fillets and flour them. Fry in butter in a frying pan.

This recipe has not changed in three hundred and fifty years, so that for the menu any dishes that followed it would be suitable. At the lunch the salad served was excellent and original.

CAULIFLOWER SALAD WITH SHRIMPS

Boil large whole cauliflower in salted water until the flowerets are tender but no more than that. Drain and press head down in a bowl while still hot so that when cold and removed from the bowl it will keep its shape. Place cauliflower flat side down on a flat round serving dish. Place 1 lb. shelled giant shrimps between the flowerets, tails pendant. Serve apart:

SAUCE MOUSSELINE

A sauce Mousseline is made by placing in the top of a double boiler the yolks of 3 eggs, 1/4 teaspoon salt and a pinch of pepper and the same of nutmeg. Put over very hot but not boiling water. Stir constantly with a wooden spoon, particularly around the sides and at the bottom. Add in very small pieces 1/4 lb. butter. Allow each piece to melt before adding the next one. When all the butter has melted and the mixture has thickened, remove from heat and add very slowly 1 tablespoon lemon juice. When the sauce is tepid add 1/2 cup whipped cream. Stir gently and serve.

This sauce is suitable with cold fish, asparagus and shell-fish salads.

At the home of a French friend, the wife of a painter not at that time celebrated—and consequently in modest circumstances—cooked plain and tasty food. She was one of those exceptional French women who made no excessive effort when she had guests. The table would be set with extra care, white linen instead of everyday coloured, the wedding silver would be brought forth, the glasses were of crystal, and a simple dessert was added to the usual fare. Her guests recognised that she was honouring them in these details. One of her very nice desserts was what is called in French, although it has no resemblance to our sweet,

FLOATING ISLAND

Separate 6 eggs. Beat the whites until stiff but not dry, adding a pinch of tragacanth, a powder that can be found at any good chemist's and which helps in the cooking of the whites of eggs and keeps them stiff. Add gradually, while beating, 5 tablespoons sugar. Place this in a mould, prepared by melting in it over very low heat 5 tablespoons sugar. Tip the mould in all directions so that the bottom and sides are completely covered. Place the beaten whites of eggs in this, gently tapping so that there are no air pockets. Put the mould uncovered in a recipient of hot water over low flame. The water should simmer but not boil. When a coarse darning needle—kept in the kitchen for such use—stuck in the centre comes out dry, remove from flame and water. When cold, carefully turn out into a hollow dish. Stir the yolks of the 6 eggs with 1/2 cup sugar until lemon coloured. Add gradually 3 cups hot milk. Pour into a saucepan and stir with a wooden spoon over very low flame until the spoon is well coated. Remove from flame and add 1 teaspoon vanilla extract and a few drops of extract of bitter almonds. Stir from time to time until cold. Pour around the caramelised whites of eggs just before serving.

Friends (whom we had not met) of friends once asked us to one of their well-known Sunday lunches, well-known because of the superlative cooking and the agglomeration of their guests. We were indeed a numerous and disparate company—the host a well-known international lawyer, a celebrated Italian cinema actress and her manager, a famous portrait painter, an American publisher and the rest of us of much lesser prominence. The house was an ancient royal hunting lodge in a large property some thirty miles from Paris. The menu and cooking were suitable to the beauty of the surroundings. The partridges had been shot by the host and his house guests, the vegetables, fruits and flowers came from the conservatories, the butter, cream and eggs from the dairy, the caviare from Russia—the iron curtain was just descending—and the wine from France, Germany and Hungary. But the fish course was French.

FILLETS OF SOLE WITH LOBSTER SAUCE

Wash the fillets of 2 soles weighing about 3/4 lb. each, reserving the rest of the fish, that is, the heads, tails, fins and small side bones. Roll the fillets and skewer them. Bring to a boil in their juice 12 fine oysters. Remove from fire and reserve the juice. Bring to a boil without water in a covered saucepan 1 quart mussels. As soon as the shells open remove from heat. Remove mussels from shells, and reserve juice. Cook covered in a small saucepan 1/2 lb. fresh mushrooms in 1 tablespoon butter, 1 tablespoon water and 1 tablespoon lemon juice. Boil for 10 minutes, reserve juice. Bring to a boil covered 1 medium-sized carrot, 1 medium-sized onion, a bouquet of 1 stalk celery, 1 twig thyme, 1 laurel leaf, 1 clove, 1 cup dry white wine and 1 cup water. Add salt and pepper, the heads, tails and bones of the fish and the juice of the mussels, oysters and mushrooms. Simmer for 1/2 hour. Remove from heat and strain. This is a *court-bouillon.*

Shell 1/2 lb. shrimps, put aside with mussels, oysters and mushrooms. Put the *court-bouillon* in covered saucepan, bring to a boil. Put in the rolled fillets, simmer for 10 minutes. Put 3 tablespoons butter in a saucepan, add 2 tablespoons crushed lobster eggs pounded through a fine sieve with potato masher. Moisten with a little of the *court-bouillon* in which the fillets have cooked. Pour into the *court-bouillon*, whip with a whisk. Stir the yolks of 3 eggs and add 1 cup hot cream. Pour into *court-bouillon* over low flame, stir, do not allow to boil. Add oysters, mussels, mushrooms and shrimps. Add in very small quantities 1/4 lb. butter. Do not stir to mix but tip saucepan in all directions. Remove from flame, add a squeeze of lemon juice and serve.

This is a royal dish.

It was a lunch to be remembered.

Another time we were invited by friends (whom we had never met) of a friend to dine on a mountain top near Lyon. This time it was at an ancient fortified farmhouse. If the company was more numerous than the one at the lunch party just described, it was more homogeneous. The dinner was composed entirely of food special to Lyon. The meat course was

ROLLED SLICES OF BEEF IN CREAM

Bone 1/4 lb. anchovies. Chop very fine 3 stalks parsley and 1 medium-sized onion, and crush 1 clove of garlic. Mash these ingredients to a paste. Cut 1 1/2 lbs. fillet of beef into four slices and spread this anchovy paste on them. Roll these slices and tie together. Put 4 tablespoons butter in iron pot over medium heat. When hot, place the rolled slices of beef in the pot. Brown lightly on all sides, add 2 cups hot veal stock, add 1/2 teaspoon pepper and simmer over low flame for 1 hour. After 3/4 hour add 3/4 cup cream in which 1 tablespoon lemon juice has been mixed. Cook for 1/4 hour further. Remove the strings and serve with

GOURMET'S POTATOES

Peel 2 lbs. potatoes and cut as for shoestring potatoes. Butter generously a mould with a tight-fitting cover. Butter a piece of paper cut to fit the bottom of mould. On this place a layer of potatoes, a sprinkling of salt and of chopped truffle previously cooked in white wine, allowing two truffles for the dish. Melt 3/4 cup butter for the dish sprinkling each layer generously. Fill the mould with layers of potatoes, salt, truffles and butter until full. Butter the cover of mould and put in preheated 350° oven for 1 hour. After 1/2 hour, lift the cover to see if the potatoes are dry. If they are, add 2 or 3 tablespoons melted butter. Continue to watch the potatoes are not dry until they are done. Carefully turn out of mould.

We did not always eat so well in French homes. Our landlady had lately married an Army officer who was a professor of map and chart making at St. Cyr, one of the historic military schools. They were living in a simple little house which was however furnished with priceless heirlooms. It was a rare pleasure to see them. Alas the lunch was not such a pleasure. The service was disorderly and careless, and a fine old Sèvres plate of the set being used for the meal was broken. The omelette was burned—a unique omelette indeed—the mutton chops were underdone, and the cream in the dessert was sour. The conversation and the wine, which had been inherited, redeemed the wretched cooking. Many years later we happened to meet the godfather of our young host at St. Cyr. He and his wife asked us to lunch to efface the memory of the disaster of the last one. It was unpretentious but exquisitely cooked. He had been stationed in Dumas where his wife had learned to cook some of the regional dishes. We had

KOBBÉ

Grind 4 cups lean mutton three times in the meat chopper. Add 1 teaspoon saffron, 1 teaspoon pepper, 1 teaspoon salt and 5 cups coarse corn meal. Mix these ingredients thoroughly with 1/2 cup oil of sesame or other bland oil. In a frying pan heat 1/2 cup oil and add 10 chopped onions and 1/2 cup pine nuts. Brown well but do not burn. Oil a fire-proof earthenware dish. Put a layer of a little less than one-third of the meat and corn-meal mixture into the dish, flatten it with the back of an oiled tablespoon or the palm of your hand. Flatten the browned onions and nuts on this layer. Cover with the rest of the meat mixture. Flatten and pour over it 4 tablespoons oil. Place in preheated 425° oven for 40 minutes. If the meat dries, add more oil, 1 table-spoon of oil at a time.*

It was explained that corn meal was the substitute purchasable in France that approached nearest the native wheat. In any case it made a succulent savoury dish. The pine nuts, saffron and sesame oil gave the dish an original flavour.

This lunch compensated largely for the one served at the godson's table.

The best foreign cooking is in the homes of the French who have been forced for one reason or another to live in their colonies. Even Italian dishes are rarely well prepared by the French. As for American dishes, they are scarcely recognisable. But after two or three years in Indo-China or Africa they return not only with the recipes of the local cooking but with the materials unobtainable in France and a knowledge of how to prepare them. This is one we ate in several different homes.

* *Note.* Sesame is an East Indian annual plant cultivated for its seeds which give a delicate faintly sweet oil much used in African and Oriental dishes. Peanut oil is a fair substitute.

MUTTON CROQUETTES (ALGERIA)

Put through meat chopper three times 4 cups lean mutton. Chop 4 onions very fine, mix well, add 2 raw eggs, 1 teaspoon salt, 1 teaspoon Spanish pepper (purchasable in any Spanish and some other grocery stores) and 1/2 teaspoon powdered cumin. Mould in floured palms of hands into round flat *croquettes*. Fry in oil over medium flame for 6 minutes on each side.

VEGETABLES AND EGGS

Skin 2 sweet peppers, remove seeds and chop. Chop 4 tomatoes, skins removed, 2 medium-sized unpeeled Italian squash. Add 2 peeled egg plants cut in thick slices. Place these vegetables in frying pan with 1/4 cup olive oil over medium flame with 1/2 teaspoon salt, and 1 clove of mashed garlic. Stir with a wooden spoon, do not allow to burn. After 1/4 hour add 4 eggs beaten as for an omelette. Remove from heat, turn into well-oiled fireproof earthenware dish and place in preheated 425° oven just long enough to set the eggs.*

This is a pleasant change from an omelette with only one of these vegetables.

These are delicious fried cookies as made in Tunis

FRIED COOKIES

Mix 4 1/3 cups flour, 1/2 cup sugar and a pinch of salt. Beat as for an omelette 3 eggs. Add slowly to dry mixture and add only enough water to hold the mixture together. If the eggs are large it may not be necessary to add any water. Mix until the dough is smooth. Roll out on a floured board to about 1/10 of an inch thickness. Cut with cookie cutter or simply in squares with a knife.

* *Note.* Italian squash (zucchini) are a sort of small vegetable marrow.

Fry in very hot deep oil only long enough to slightly colour them. Remove with perforated flat spoon. Place on absorbent paper and as quickly as possible paint with pastry brush one side of the cookies with honey and sprinkle generously with not too finely chopped pine nuts.

These Colonial dishes add variety to what are frequently in middle-class French families well-cooked but monotonous menus. Often either the master or the mistress of the household comes from a different province from the other, and this gives more variety to their menus. And then there are the dishes that over the years have been introduced by various cooks. Friends have told me that this has compensated for the annoyance and trouble of changing cooks.

At the country home of friends in Burgundy we ate several delicacies. To our great delectation we had frogs' legs twice in three days, the first time

FRIED FROGS' LEGS

Marinate for an hour 100 frogs' legs in 1 cup olive oil and 1 teaspoon salt, turning frequently. Drain well and wipe dry. Cover them with this

BATTER FOR FRYING

Separate 2 eggs. Stir the yolks, add a pinch of salt and 2 tablespoons olive oil. When well mixed, add slowly, stirring with a wooden spoon, 1 cup and 1 tablespoon flour. Beat the whites of eggs and add them to mixture. Put aside for an hour before using.

And then,

FROGS' LEGS WITH CREAM

Put 100 frogs' legs with 1/4 cup butter into frying pan over medium heat, add 1/2 teaspoon salt and 1/4 teaspoon pepper. Stir them with wooden spoon until browned. Remove from pan, place on preheated serving dish. Keep hot. Add to frying pan 1 cup hot heavy cream mixed with 1/2 cup *Béchamel* sauce. Mix with butter in pan, stirring sides and bottom. Add 4 tablespoons butter. Do not allow to boil. Stir until melted, strain and pour over frogs' legs. Sprinkle with 1 tablespoon finely chopped parsley. Serve at once.

This is the best way to cook frogs' legs.

Many years ago we had an unusual lunch in an unusual setting. Gertrude Stein and I were asked to join some friends at a property in the Camargue, a peninsula of some twenty-five square miles in the delta of the Rhône. We drove down one morning in the late autumn into the deserted marshlands over bridges of boats to the Domaine of S, where we were to meet and lunch. The house was old and not lived in except for a guardian. It was used by its owner when he and his friends went shooting and fishing in the neighbourhood. The men would be bringing back fish and game (the French do not care for it high) that we would eat for lunch. In the huge room a blazing log fire roared. The women were setting the table and unwrapping the prepared dishes they had brought with them, meat and chicken *pâtés* which would be heated, aspics and butter and eggs, glasses, silver and linen. Gertrude Stein and I were suddenly taken out of doors to see two flamingoes drinking and some small white bulls, descendants of wild ones. There was a warm noisy welcome for the men when they returned, the bags were emptied and the birds and fish chosen for cooking. They were given to the guardian to pluck and clean, with some supervision from one woman of us at a time. The fire was very quickly reduced to one suitable for roasting on a spit. The mallards which had been chosen would not take long. The spit would take eight at a time, and there would be a second roasting during the carving and eating of the first. The lampreys were skinned and cleaned—the men were very proud of having found them at that season in the nearby Rhône—and they were cut in long pieces, each one wrapped in a thin slice of fat back of pork, and grilled on charcoal while this sauce was being prepared in an immensely ancient silver chafing dish.

SAUCE FOR LAMPREYS OR OTHER GRILLED FISH

For 1 3/4 lbs. fish, melt 4 tablespoons butter in a saucepan, add 1 tablespoon flour. Stirring constantly, slowly add 1 cup hot water, 1 cup Madeira or port, 1/2 lb. chopped mushrooms, 1 small chopped onion, 1/4 teaspoon salt, 3 crushed cloves of garlic, 6 whole peppers, 1 small sprig of thyme and 1 stalk of basil. Simmer until it is reduced by half. Strain and skim.

The straining and skimming were omitted with the general informality that prevailed.

The ducks, carefully basted with the butter from the dripping pan and their juice dripping from them into it, were cooked longer by a quarter of an hour than if roasted in an oven. They would be cooked longer than Gertrude Stein and I cared for them to be. The French dislike to see red juice when game is carved, they prefer their mutton underdone. The fish and the ducks were cooked to perfection. It was, with the endless dishes preceding and following the two courses, a Gargantuan feast. The guests drove to their homes in various directions, but a group of us went to spend the night at the inn at Les Baux to see the Camargue in the moonlight. Providently in my handbag was a little jar of American powdered coffee, which was a blessing for our breakfast and a novelty to our French friends.

There have been many more of these invitations to meals in French homes of which the greater number so much resembled each other that in a very short time they became indistinguishable. And this is not a reproach from a guest. It is a quality that limits them to what the French consider suitable—and that is their ideal. They achieve a harmony, experience in their preparation and a justifiable pride in the finely balanced French cooking.

III.

DISHES FOR ARTISTS

BEFORE COMING TO PARIS I WAS INTERESTED IN FOOD BUT NOT IN doing any cooking. When in 1908 I went to live with Gertrude Stein at the rue de Fleurus she said we would have American food for Sunday-evening supper, she had had enough French and Italian cooking; the servant would be out and I should have the kitchen to myself. So I commenced to cook the simple dishes I had eaten in the homes of the San Joaquin Valley in California—fricasseed chicken, corn bread, apple and lemon pie. Then when the pie crust received Gertrude Stein's critical approval I made mince-meat and at Thanksgiving we had a turkey that Hélène the cook roasted but for which I prepared the dressing. Gertrude Stein not being able to decide whether she preferred mushrooms, chestnuts or oysters in the dressing, all three were included. The experiment was successful and frequently repeated; it gradually entered into my repertoire, which expanded as I grew experimental and adventurous.

BASS FOR PICASSO

One day when Picasso was to lunch with us, I decorated a fish in a way that I thought would amuse him. I chose a fine striped bass and cooked it according to a theory of my grandmother who had no experience in cooking and who rarely saw her kitchen but who had endless theories about cooking as well as about many other things. She contended that a fish having lived its life in water, once caught, should have no further contact with the element in which it had been born and raised. She recommended that it be roasted or poached in wine or cream or butter. So I made a *court-bouillon* of dry white wine with whole peppers, salt, a laurel leaf,* a sprig of thyme, a blade of mace, an onion with a clove stuck in it, a carrot, a leek and a bouquet of *fines herbes.* This was gently boiled in the fish-kettle for 1/2 hour and then put aside to cool. Then the fish was placed on the rack, the fish-kettle covered and slowly brought to a boil and the fish poached for 20 minutes. Taken from the fire it was left to cool in the *court-bouillon.* It was then carefully drained, dried and placed on the fish platter. A short time before serving it I covered the fish with an ordinary mayonnaise and, using a pastry tube, decorated it with a red mayonnaise, not coloured with catsup—horror of horrors—but with tomato paste. Then I made a design with sieved hard-boiled eggs, the whites and the yolks apart, with truffles and with finely chopped *fines herbes.* I was proud of my chef d'oeuvre when it was served and Picasso exclaimed at its beauty. But, said he, should it not rather have been made in honour of Matisse than of me.

Picasso was for many years on a strict diet; in fact he managed somehow to continue it through the World War and the Occupation and, characteristically, only relaxed after the Liberation. Red meat was proscribed but that presented no difficulties for in those days beef was rarely served by the French except the inevitable roast fillet of beef with *sauce Madère.* Chicken too was not well considered, though a roast leg of mutton was viewed with more favour. Or we would have a tender loin of veal preceded by a spinach *soufflé*, spinach having been highly recommended by Picasso's doctor and a *soufflé* being the least objectionable way of preparing it. Could it not be made more interesting by adding a sauce. But what sauce would Picasso's diet permit. I would give him a choice. The *soufflé* would be cooked in a well-buttered mould, placed in boiling water and when sufficiently cooked turned into a hollow dish around which in equal divisions would be placed a Hollandaise sauce, a cream sauce and a

* *Note.* The leaf must come from Apollo's Laurel (*Laurus Nobilis*), better known outside France as the bay.

tomato sauce. It was my hope that the tri-coloured sauces would make the spinach *soufflé* look less nourishing. Cruel enigma, said Picasso, when the *soufflé* was served to him.

The only painter who ever gave me a recipe was Francis Picabia and though it is only a dish of eggs it merits the name of its creator.

OEUFS FRANCIS PICABIA

Break 8 eggs into a bowl and mix them well with a fork, add salt but no pepper. Pour them into a saucepan—yes, a saucepan, no, not a frying pan. Put the saucepan over a very, very low flame, keep turning them with a fork while very slowly adding in very small quantities 1/2 lb. butter—not a speck less, rather more if you can bring yourself to it. It should take 1/2 hour to prepare this dish. The eggs of course are not scrambled but with the butter, no substitute admitted, produce a suave consistency that perhaps only *gourmets* will appreciate.

When the Germans in 1940 were advancing we were at Bilignin and had no precise information concerning their progress through France. Could one believe the radio. We didn't. We heard cannon-fire. Then it grew louder. The next morning dressing at the window I saw German planes firing on French planes, not more than two miles away. This decided me to act in the way any forethoughtful house-keeper should. We would take the car into Belley and make provision for any eventuality as I had done that April morning of 1906 when the fire in San Francisco had broken out after the earthquake. Then I had been able to secure two hams and my father had brought back four hundred cigarettes. With these one might, he said, not only exist but be able to be hospitable. So at Belley we bought two hams and hundreds of cigarettes and some groceries—the garden at Bilignin would provide fruit and vegetables. The main road was filled with refugees, just as it had been in 1914 and in 1917. Everything that was happening had already been experienced, like a half-awakening from nightmare. The firing grew louder and then the first armoured car flew past. Crushed, we took the little dust road back to Bilignin. The widow Roux, who for many summers had been our devoted servant and later during the Occupation proved to be our loyal friend, opened the big iron gates to let the car through and we unloaded the provisions. What were we to do with the two enormous uncooked hams. In what could we cook them and in what way so that they would keep indefinitely. We decided upon Eau-de-Vie de Marc for which the Bugey is well known. It seemed madly extravagant but we lived on those two hams during the long lean winter that followed and well into the following spring, and the Eau-de-Vie de Marc in which they were cooked, carefully bottled and corked, toned up winter vegetables. We threw nothing, but absolutely nothing away, living through a war in an occupied country.

The Baronne Pierlot, our neighbour, was *châtelaine* at Beon, some ten miles away. One day, before the war, we had driven over to a *gouté** to which she had bidden us. It was being served in the summer dining-room whose windows and door gave on to a vast terrace. In the foreground was the marsh of the Rhône Valley lately reclaimed by the planting of Lombardy poplars, to the south the mountains of the Grande Chartreuse, to the left in the distance the French Alps and over it all the Tiepolo blue sky. The table in the dining-room set for twenty or more was elaborately decorated with pink roses. Madame Pierlot's observant eye passed quickly and lightly over each object on the table. I heard her tell the *valet-de-chambre* to ask the cook for the *pièce-de-résistance* and to place it in the empty space waiting for it in the centre of the table. But Marc did not leave the room, he merely took a cake from the serving table and put it in the empty space. There was evidently some *contretemps*. I was enlightened when I caught knowing looks pass-

* *Note.* Here, a lavish afternoon tea-party.

ing between Gertrude Stein and one of the daughters-in-law of the house. It was Gertrude Stein's white poodle, a very neat thief, who had done away with whatever had been in the centre of the table. Later when Madame Pierlot, to show that she had forgiven the dog, threw him a piece of cake we could not protest that it was against our principles to reward a misdeed.

Madame Pierlot was an old friend of Paul Claudel and there had been a long controversial correspondence over the years, largely on religious subjects; Claudel a devout Catholic, Madame Pierlot not. Bernard Faÿ said that she had been converted once and forever by Jean-Jacques Rousseau. She told Gertrude Stein one day that Claudel's letters were beginning to bore her and she was equally bored by having to answer him. She had written to him saying that they would no longer defend their opinions, that they would no longer write to each other, but they would remain the same good old friends they had always been. Claudel could not resist having the last word. He wrote that in spite of her continuously avowed unbelief he was certain that when he died he would find her in Heaven welcoming him with arms extended, to which she replied—Who tells you that I am to die before you.

If Madame Pierlot was known as an exquisite hostess it was not only for her wit and charm or for her impeccable taste in choosing her guests and her menus, but also for the care with which her old cook, Perrine, prepared the menus. Madame Pierlot told me that when she was engaging her to come to be her cook she asked her if she knew how to prepare several complicated dishes which she mentioned. She saw that Perrine had had a large experience. As she was well recommended, I decided, Madame Pierlot told me, to engage her, but I told her that it was on the condition that she would forget everything she knew and follow the recipes and the instructions I would give her.

Our enchanting old friend was as original in her housekeeping as in everything else. Long ago the *Figaro* which was then the newspaper read by the fashionable world asked well-known society women to contribute recipes which were to be printed in a special column. When Madame Pierlot was asked to be one of the contributors she sent the recipe for

GIGOT DE LA CLINIQUE

A surgeon living in the provinces, as fond of good cheer as he was learned, invented this recipe which we acquired by bribing his cook. No leg of venison can compare with a simple leg of mutton prepared in the following manner. Eight days in advance you will cover the leg of mutton with the marinade called Baume Samaritain, composed of wine—old Burgundy, Beaune or Chambertin—and virgin olive oil. Into this balm to which you have already added the usual condiments of salt, pepper, bay leaf, thyme, beside an atom of ginger root, put a pinch of cayenne, a nutmeg cut into small pieces, a handful of crushed juniper berries and lastly a dessertspoon of powdered sugar (effective as musk in perfumery) which serves to fix the different aromas. Twice a day you will turn the *gigot.* Now we come to the main point of the preparation. After you have placed the *gigot* in the marinade you will arm yourself with a surgical syringe of a size to hold 1/2 pint which you will fill with 1/2 cup of cognac and 1/2 cup of fresh orange juice. Inject the contents of the syringe into the fleshy part of the *gigot* in three different spots. Refill the syringe with the same contents and inject into the *gigot* twice more. Each day you will fill the syringe with the marinade and inject the contents into the *gigot.* At the end of the week the leg of mutton is ready to be roasted; perfumed with the condiments and the spices, completely permeated by the various flavours, it has been transfused into a strange and exquisite venison. Roast and serve with the usual venison sauce to which has been added just before serving 2 tablespoons of the blood of a hare.*

* *Note.* A marinade is a bath of wine, herbs, oil, vegetables, vinegars and so on, in which fish or meat destined for particular dishes repose for specified periods and acquire virtue.

Everyone thought that the syringe was a whimsy, that Madame Pierlot was making mock of them. Not at all. Years later I found it in that great collection of French recipes, Bertrand Guegan's *Le Grand Cuisinier Français.* The Baronne Pierlot's recipe is classified, it has entered into the *Grande Cuisine Française.*

IV.

MURDER IN THE KITCHEN

COOK-BOOKS HAVE ALWAYS INTRIGUED AND SEDUCED ME. WHEN I WAS still a dilettante in the kitchen they held my attention, even the dull ones, from cover to cover, the way crime and murder stories did Gertrude Stein.

When we first began reading Dashiell Hammett, Gertrude Stein remarked that it was his modern note to have disposed of his victims before the story commenced. Goodness knows how many were required to follow as the result of the first crime. And so it is in the kitchen. Murder and sudden death seem as unnatural there as they should be anywhere else. They can't, they can never become acceptable facts. Food is far too pleasant to combine with horror. All the same, facts, even distasteful facts, must be accepted and we shall see how, before any story of cooking begins, crime is inevitable. That is why cooking is not an entirely agreeable pastime. There is too much that must happen in advance of the actual cooking. This doesn't of course apply to food that emerges stainless from deep freeze. But the marketing and cooking I know are French and it was in France, where freezing units are unknown, that in due course I graduated at the stove.

In earlier days, memories of which are scattered among my chapters, if indulgent friends on this or that Sunday evening or party occasion said that the cooking I produced wasn't bad, it neither beguiled nor flattered me into liking or wanting to do it. The only way to learn to cook is to cook, and for me, as for so many others, it suddenly and unexpectedly became a disagreeable necessity to have to do it when war came and Occupation followed. It was in those conditions of rationing and shortage that I learned not only to cook seriously but to buy food in a restricted market and not to take too much time in doing it, since there were so many more important and more amusing things to do. It was at this time, then, that murder in the kitchen began.

The first victim was a lively carp brought to the kitchen in a covered basket from which nothing could escape. The fish man who sold me the carp said he had no time to kill, scale or clean it, nor would he tell me with which of these horrible necessities one began. It wasn't difficult to know which was the most repellent. So quickly to the murder and have it over with. On the docks of Puget Sound I had seen fishermen grasp the tail of a huge salmon and lifting it high bring it down on the dock with enough force to kill it. Obviously I was not a fisherman nor was the kitchen table a dock. Should I not dispatch my first victim with a blow on the head from a heavy mallet? After an appraising glance at the lively fish it was evident he would escape attempts aimed at his head. A heavy sharp knife came to my mind as the classic, the perfect choice, so grasping, with my left hand well covered with a dishcloth, for the teeth might be sharp, the lower jaw of the carp, and the knife in my right, I carefully, deliberately found the base of its vertebral column and plunged the knife in. I let go my grasp and looked to see what had happened. Horror of horrors. The carp was dead, killed, assassinated, murdered in the first, second and third degree. Limp, I fell into a chair, with my hands still unwashed reached for a cigarette, lighted it, and waited for the police to come and take me into custody. After a second cigarette my courage returned and I went to prepare poor Mr. Carp for the table. I scraped off the scales, cut off the fins, cut open the underside and emptied out a great deal of what I did not care to look at, thoroughly washed and dried the fish and put it aside while I prepared

CARP STUFFED WITH CHESTNUTS

For a 3-lb. carp, chop a medium-sized onion and cook it gently in 3 tablespoons butter. Add a 2-inch slice of bread cut into small cubes which have previously been soaked in dry, white wine and squeezed dry, 1 tablespoon chopped parsley, 2 chopped shallots,

1 clove of pressed garlic, 1 teaspoon salt, 1/4 teaspoon freshly ground pepper, 1/4 teaspoon powdered mace, the same of laurel (bay) and of thyme and 12 boiled and peeled chestnuts. Mix well, allow to cool, add 1 raw egg, stuff the cavity and head of the fish, carefully snare with skewers, tie the head so that nothing will escape in cooking. Put aside for at least a couple of hours. Put 2 cups dry white wine into an earthenware dish, place the fish in the dish, salt to taste. Cook in the oven for 20 minutes at 375°. Baste, and cover the fish with a thick coating of very fine cracker crumbs, dot with 3 tablespoons melted butter and cook for 20 minutes more. Serve very hot accompanied by noodles. Serves 4. The head of a carp is enormous. Many continentals consider it the most delectable morsel.

NOODLES

Sift 2 cups flour, 1 teaspoon salt and a pinch of nutmeg, add the yolks of 5 eggs and 1 whole egg. Mix thoroughly with a fork and then knead on a floured board, form into a ball, wrap in a cloth and put aside for several hours. Divide into three parts. Roll each one in turn on a lightly floured board to tissue-paper thinness. Dry for 1/2 hour, roll up and cut into strips 1/4 inch wide. Bring 1 quart water with 1 teaspoon salt to a hard boil. Place noodles a few at a time into boiling water, stir gently with a fork, reduce heat and boil slowly for 10 minutes. Drain off all the water and add 3 tablespoons melted butter. These noodles are very delicate. Serves 4.

It was in the market of Palma de Mallorca that our French cook tried to teach me to murder by smothering. There is no reason why this crime should have been committed publicly or that I should have been expected to participate. Jeanne was just showing off. When the crowd of market women who had gathered about her began screaming and gesticulating, I retreated. When we met later to drive back in the carry-all filled with our marketing to Terreno where we had a villa I refused to sympathise with Jeanne. She said the Mallorcans were bloodthirsty, didn't they go to bullfights and pay an advanced price for the meat of the beasts they had seen killed in the ring, didn't they prefer to chop off the heads of innocent pigeons instead of humanely smothering them which was the way to prevent all fowl from bleeding to death and so make them fuller and tastier. Had she not tried to explain this to them, to teach them, to show them how an intelligent humane person went about killing pigeons, but no they didn't want to learn, they preferred their own brutal ways. At lunch when she served the pigeons Jeanne discreetly said nothing. Discussing food which she enjoyed above everything had been discouraged at table. But her fine black eyes were eloquent. If the small-sized pigeons the island produced had not achieved jumbo size, squabs they unquestionably were, and larger and more succulent squabs than those we had eaten at the excellent restaurant at Palma.

Later we went back to Paris and then there was war and after a lifetime there was peace. One day passing the *concierge's loge* he called me and said he had something someone had left for us. He said he would bring it to me, which he did and which I wished he hadn't when I saw what it was, a crate of six white pigeons and a note from a friend saying she had nothing better to offer us from her home in the country, ending with But as Alice is clever she will make something delicious of them. It is certainly a mistake to allow a reputation for cleverness to be born and spread by loving friends. It is so cheaply acquired and so dearly paid for. Six white pigeons to be smothered, to be plucked, to be cleaned and all this to be accomplished before Gertrude Stein returned for she didn't like to see work being done. If only I had the courage the two hours before her return would easily suffice. A large cup of strong black coffee would help. This was before a lovely Brazilian told me that in her country a large cup of black coffee was always served before going to bed to ensure a good night's rest. Not yet having acquired this knowledge the black coffee made me lively and courageous. I carefully found the spot on poor innocent Dove's throat where I was to press and pressed. The realization had never come to me before that one saw with one's fingertips as well as one's eyes. It was a most unpleasant experience, though as I laid out one by one the sweet young corpses there was no denying one could become accustomed to murdering. So I plucked the pigeons, emptied them and was ready to cook

BRAISED PIGEONS ON CROÛTONS

For 6 pigeons cut 1/2 lb. salt pork in small cubes, place in Dutch oven with 6 tablespoons butter, place pigeons in oven, brown slightly, cover and cook over low flame for 1 hour turning and basting frequently. While pigeons are cooking wash and carefully dry 2 lbs. mushrooms. Chop them very fine, and pass through a coarse sieve, cook over brisk fire in 1/4 lb. butter until liquid has evaporated. Reduce flame and add 1 cup heavy cream sauce and 1/2 cup heavy cream. Spread on 6 one-half-inch slices of bread that have been lightly browned in butter. Spread the *purée* of mushrooms on the *croûtons*. Place the pigeons on the *croûtons*. Skim the fat from the juice in the Dutch oven, add 2 tablespoons Madeira, bring to a boil and pour over pigeons. Salt for this dish depends upon how salty the pork is. Serves 6 to 12 according to size of pigeons.

The next murder was not of my doing. During six months which we spent in the country we raised Barbary ducks. They are larger than ordinary ducks and are famous for the size of their livers. They do not quack and are not friendly. Down in the Ain everyone shoots. Many of the farmers go off to work in the fields with a gun slung over a shoulder and not infrequently return with a bird or two. Occasionally a farmer would sell us a pheasant or a partridge. An English friend staying with us, astonished to find farmers shooting, remarked, When everyone shoots no one shoots. Our nearest neighbour had a

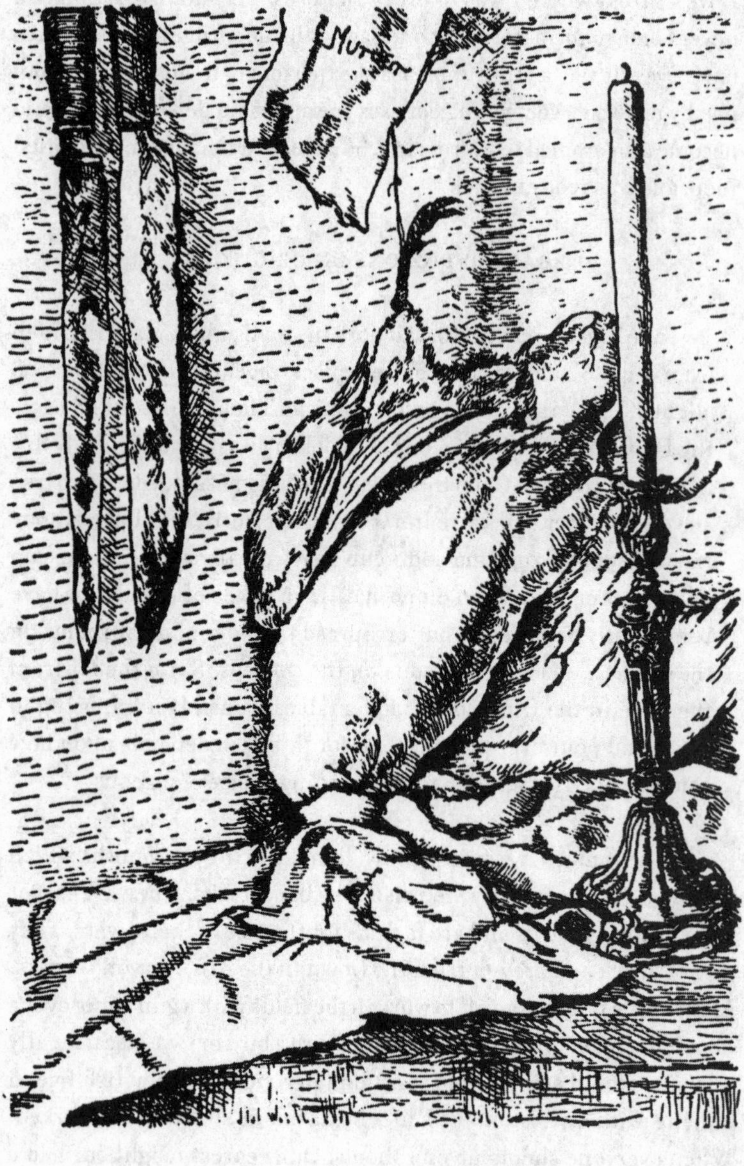

so-called bird dog, mongrel she certainly was, ruby coat like an Irish setter but her head was flat, her paws too large, her tail too short. We would see Diane on the road, she was not sympathetic. The large iron portals at Bilignin were sometimes left open when Gertrude Stein took the car out for a short while, and one morning Diane, finding them open, came into the court and saw the last of our Barbary ducks, Blanchette, because she was blue-black. Perhaps innocently perhaps not, opinion was divided later, she began to chase Blanchette. She would come running at the poor bewildered duck from a distance, charge upon her, retreat and recommence. The cook, having seen from the kitchen window what was happening, hastened out. The poor duck was on her back and Diane was madly barking and running about. By the time I got to the court the cook was tenderly carrying a limp Blanchette in her arms to the kitchen. Having chased Diane out of the court, I closed the portals and returned to my work in the vegetable garden supposing the episode to be over. Not at all. Presently the cook appeared, her face whiter than her apron. Madame, she said, poor Blanchette is no more. That wretched dog frightened her to death. Her heart was beating so furiously I saw there was but one thing to do. I gave her three tablespoonsfuls of *eau-de-vie*, that will give her a good flavour. And then I killed her. How does Madame wish her to be cooked. Surprised at the turn the affair had taken, I answered feebly, With orange sauce.

There was considerable talk in the hamlet. While we were walking along the road someone would say What a pity, or Your beautiful bird! to which we would answer that we would have had to be eating her soon anyway. But Diane's master did not know what attitude to take until I sent his wife a basket of globe egg plants, almost white and yellow tomatoes and a few gumbos (okra), none of which she had seen before. Then he came to thank us for his wife and presented a large pot of fresh butter she had sent us. He knew our cook felt that his dog had caused the death of our duck. We wiped out the memory of the misadventure in thanking each other for the gifts. So Blanchette was cooked as

DUCK WITH ORANGE SAUCE

Put the bird aside and cook the rest of the giblets including the neck in 2 cups water with 1 teaspoon salt, 1/4 teaspoon pepper, 1 small onion with a clove stuck in it, a shallot, 1/2 laurel leaf, a sprig of thyme and a small blade of mace. Cover and cook slowly. When the juice has reduced to 1 cup put aside. Cut 1 peeled orange into half a dozen pieces and put inside the duck. Cut the orange peel into small pieces and boil covered in 1/2 cup water for 10 minutes. Roast the duck in a 400° oven in a pan with 3 tablespoons butter for 1/2 hour, basting and turning the duck three times. Put the orange peel and the liver in a mortar. Moisten with 1/3 cup of the best white curaçao and crush to an even paste. Add to this the cup of giblet juice and the juice in the pan from which the fat has been skimmed. Heat thoroughly but do not allow to boil, strain and serve in preheated metal sauce boat. Place very thinly sliced unpeeled oranges on the duck and serve. Sufficient for 4.

Many times I held the thought to kill a stupid or obstinate cook, but as long as the thought was held murder was not committed. Then a gay and enchanting Austrian came to cook for me. He was a perfect cook. Quietly and expeditiously, Frederich, as I shall call him, prepared the most intricate and complicated dishes for us, nothing was too much trouble for him to undertake. He would make us ice cream in individual moulds in the form of eggs on a nest of coloured spun sugar. He delighted in making cakes that represented objects appropriate to each person, a book for Gertrude Stein, a rose for Sir Francis Rose, a peacock for a very vain young lady and a little dog for me. He used to receive the visits of an extremely pretty young girl, Duscha, who looked as if she had stepped out of an Offenbach opera. Gertrude Stein and I were delighted with them. At Christmas we asked them to accept amongst their gifts a supper with champagne at the restaurant of their choice for the traditional *reveillon.* Gradually Frederich began to confide in me. Life was not as happy for him as it had been. In the beginning there was only his fiancée Duscha, his angel, but now there was a second, a devil, who wanted him to marry her and who was threatening to kill him if he didn't. And he told us that he and Hitler had been born in the same village and that anyone in the village was like all the others and that they were all a little strange. This was in 1936 and we already knew Hitler was very strange indeed. Frederich was perhaps not so much strange as weak, loving wine, women and song. But he continued to be a perfect cook. He had been for several years a cook at Frau Sacher's restaurant and frequently baked us the well-known

SACHER TORTE

Cream 1/2 cup butter, gradually add 1 cup sugar, the grated peel of 1 lemon, 4 ozs. melted chocolate, the yolks of 6 eggs, fold in the beaten whites of 6 eggs and 3 tablespoons flour. Butter and flour a flat cake pan and bake for 40 minutes in a 325° oven. Let cool in pan. When perfectly cold, cut in half and spread the following mixture between the two layers.

2 ozs. chocolate melted, to which add 1 teaspoon powdered coffee dissolved in 1/2 cup hot water. When perfectly smooth beat in 2 yolks of eggs. Beat 1 cup heavy cream sweetened with 3 tablespoons icing sugar. Add first mixture to the whipped cream.

Cover the cake with apricot jelly or strained apricot jam and ice with chocolate icing.

Frederich also liked to serve

LINZER TORTE

1/2 cup powdered almonds, 1 cup flour, 1/2 cup butter, 1/2 cup sugar, the yolks of 2 boiled eggs sieved, a pinch of cinnamon, a pinch of nutmeg, grated peel of 1/2 lemon. Cut the butter and flour with knives or pastry blender, add the other ingredients in the above order, finally add 3 teaspoons rum. Put aside in the refrigerator for a couple of hours. Roll out three-quarters of the dough and fit into buttered pie plate with detachable bottom, fill with raspberry jam. Roll out rest of dough, cut with pastry wheel into strips 1/2 in. wide and place on pie in lattices. Paint lightly with beaten egg. Bake in 350° oven for 1/2 hour.

Here is the last dish Frederich served us.

GYPSY GOULASH

1 1/2 lbs. fillet of beef in slices of 1/4 inch thickness, cut in lengths of 1/4 inch width, browned in lard with 1 teaspoon salt, 1 tablespoon paprika and 1 tablespoon flour, 4 large onions sliced, 3/4 lb. potatoes sliced. When lightly browned add 2 cups red wine, 1 cup sour cream and enough *bouillon* to cover. Put in covered casserole in 375° oven for 1 hour. Add 1/2 cup sour cream before serving. Serve with noodles. Serves 4.

One afternoon as Gertrude Stein and I were coming home someone came out of our door and passed in the court. She had small snappy dark eyes. The devil, Gertrude Stein inquired. Presumably, I answered. The glimpse I had of her left me uneasy for Frederich. We wanted him to be happy and to stay with us as our servant. Later I went into the kitchen to see him. He was sitting at the table, his head in his arms. He jumped up when he saw me. What is it, I asked. The devil, Madame, the devil came to see me with a bottle of precious Tokay as a gift. The devil wanted to poison me, to kill me. The fiend poured a glass and passed it to me. Just as I was about to toast her I noticed that she had poured none for herself, her glass was empty, and that she had not taken out the cork with a corkscrew. She was going to poison me. I threw the bottle at her. I shoved her about. I threw her out. Oh Madame, the devil will get me yet or she will kill me. I sent him off to his room.

The next morning there was no Frederich in the kitchen. Towards noon I asked the *concierge* to go up to his room to see what had happened. He returned to report that the door was open, the room empty except for a strapped trunk. He had not seen Frederich all morning but the dark lady had been there about two hours ago. What could we do. Nothing but wait for Duscha to turn up which she did late in the afternoon. As pretty, dainty and as elegant as usual but her eyes red and swollen. She had had a wire from Frederich as long as a letter which proved, she said, how distraught he was. He had gone off with the devil, useless to hunt for them, they were leaving Paris. He would always love his angel but their happiness together was over forever. She should go to tell this to the good ladies, they would pay her what they owed him, for three weeks and six days, and with this she should buy herself a *frivolité* as a last souvenir of her adoring Frederich.

While I was reading this Duscha was gently sobbing into a delicate white handkerchief. I led her into the big room and left her with Gertrude Stein while I prepared tea. She came running when she saw the tray with three cups. But she put her handkerchief away and quietly drank several cups of tea and ate the last of Frederich's perfect Viennese pastry that we were to taste. What are you going to do, we asked Duscha. Go on with my work with the good kind Princess, she will understand. When my eyes are no longer red and I shall have forgotten sweet weak Frederich life will begin again. Then I paid her her faithless lover's wages. She thanked me and counted the bills. With a sob and a sigh she neatly folded and put them in her handbag. Let us hear from you, I said as she left.

But for months we didn't, then we received wedding announcements. In France this is done by the bride's family to the left, by the groom's family to the right. Duscha's family way off in an unpronounceable village in Austria had the honour and so on and then the groom's family, two grandmothers and a grandfather, his parents, his brothers, his sisters, all sprinkled with military medals and *Legions d'Honneur* and civil titles, announced that the son was marrying Duscha. She had entered a well-established *bourgeois* family with nothing more to fear.

This is the last souvenir of Frederich,

A TENDER TART

1/2 cup and 1 tablespoon butter, 1 cup and 2 tablespoons flour, 1 egg yolk, blend with knives or pastry blender, add only enough water to hold together, knead lightly, put aside in refrigerator. Stir 2 eggs and 1 cup plus 2 tablespoons sugar for 20 minutes. Do not beat. Add 1 teaspoon vanilla and 1 cup finely chopped hazel nuts. Roll out a little more than half the dough, place in deep pie plate with detachable bottom, fill with egg-sugar-nut mixture. Roll out remaining dough and cover tart, press the edges together so that the bottom and top crusts adhere. Bake for 1/2 hour in 350° oven. Exquisite.

V.

"BEAUTIFUL SOUP"

FROM MURDER TO DETECTION IS NOT FAR. AND HERE IS A NOTE ON tracking a soup to its source. It was as a result of eating *gazpacho* in Spain lately that I came to the conclusion that recipes through conquests and occupations have travelled far. After the first ineffable *gazpacho* was served to us in Malaga and an entirely different but equally exquisite one was presented in Seville the recipes for them had unquestionably become of greater importance than Grecos and Zurbarans, than cathedrals and museums. Surely the calle de las Sierpes, the liveliest, most seductive of streets, would produce the cook-book that would answer the burning consuming question of how to prepare a *gazpacho.* Down the narrow Sierpes where only pedestrians are permitted to pass, with its deluxe shops of fans, boots and gloves, toys and sweets, its smart men's clubs on either side whose members sit three tables deep sipping iced drinks and evaluating the young ladies who pass, at the end of the street was the large book shop remembered from a previous visit forty years before. Cook-books without number, exactly eleven, were offered for inspection but not a *gazpacho* in any index. Oh, said the clerk, *gazpachos* are only eaten in Spain by peasants and Americans. Choosing the book that seemed to have the fewest French recipes, I hurried back to Zurbaran and Greco, to museums and cathedrals.

At Cordoba there was another and suaver *gazpacho*, at Segovia one with a more vulgar appeal, outrageously coarse. There was nothing to do but to resign oneself to an experimental laboratory effort as soon as a kitchen was available. Upon the return from Spain my host at Cannes, a distinguished Polish-American composer, a fine *gourmet* and experienced cook, listened to the story of the futile chase for *gazpacho* recipes, for their possible ingredients. Ah, said he, but you are describing a *chłodnik*, the Polish iced soup. Before he had had time to prepare it for us a Turkish guest arrived and he hearing about the *gazpachos* and the *chłodnik* said, You are describing a Turkish *cacik*. Perhaps, said I. It was confusing. He said he would prepare a *cacik* for us. It was to be sure an iced soup, but the Turk had not the temperament of a great cook, he should not have accepted olive oil as a substitute for the blander oil of sesame. Then we had the *chłodnik*, a really great dish worthy of its Spanish cousins. But that was not the end. There was the Greek *tarata*. Yes indeed, it was confusing, until one morning it occurred to me that it was evident each one of these frozen soups was not a separate creation. Had the Poles passed the recipe to their enemy the Turks at the siege of Vienna or had it been brought back to Poland much earlier than that from Turkey or Greece? Or had it been brought back by a crusader from Turkey? Had it gone to Sicily from Greece and then to Spain? It is a subject to be pursued. Well, here are the seven Mediterranean soups.

GAZPACHO OF MALAGA (Spanish)

4 cups veal broth cooked with 2 cloves of garlic and a large Spanish onion.

1 large tomato peeled, with its seeds removed, and cut in minute cubes.

1 small cucumber peeled, with its seeds removed, and cut in minute cubes.

1/2 sweet red pepper, skin and seeds removed, cut in minute cubes.

4 tablespoons cooked rice.

2 tablespoons olive oil.

Mix thoroughly and serve ice-cold.

Sufficient for 4 though double the quantity may not be too much!

GAZPACHO OF SEVILLE

In a bowl put 4 crushed cloves of garlic.

1 teaspoon salt, 1/2 teaspoon powdered Spanish pepper.

Pulp of 2 medium-sized tomatoes crushed.

Mix these ingredients thoroughly and add drop by drop 4 tablespoons olive oil.

Add 1 Spanish onion cut in tissue-paper-thin slices.

1 sweet red or green pepper, seeds removed and cut in minute cubes.

1 cucumber peeled, seeds removed and cut in minute cubes.

4 tablespoons fresh white breadcrumbs.

Add 3 cups water, mix thoroughly.

Serve ice-cold.

GAZPACHO OF CORDOBA

2 cloves of crushed garlic, 2 cucumbers peeled, seeds removed and minutely cubed.

2 tablespoons olive oil.

2 cups water.

2 cups heavy cream.

2 teaspoons cornflour.

1 teaspoon salt.

Mix thoroughly the first three ingredients. Bring the water to a boil with the salt. Mix the cornflour with 3 additional tablespoons water, add to the boiling water. When the cornflour is cooked and the water thickened pour it over the garlic, cucumbers and oil. Let it cool and gradually add the cream. Serve ice-cold.

GAZPACHO OF SEGOVIA

4 cloves of garlic pressed.

1 teaspoon ground Spanish pepper.

1 teaspoon salt.

1/2 teaspoon cumin powder.

2 tablespoons finely chopped fresh basil or 3/4 tablespoon powdered basil.

4 tablespoons olive oil.

1 Spanish onion cut in minute cubes.

2 tomatoes peeled, seeds removed and cut in minute cubes.

2 cucumbers peeled, seeds removed and cut in minute cubes.

1 red sweet pepper, seeds removed and cut in minute cubes.

2 tablespoons fresh white breadcrumbs.

4 cups water.

Put the first six ingredients in a bowl and add drop by drop the olive oil. When this has become an emulsion add the dry breadcrumbs and the prepared onion, cucumbers and the tomatoes. Then add the water. Mix thoroughly. Serve ice-cold.

CHŁODNIK (Polish)

2 ozs. lean veal cut in small pieces cooked in water to cover.

2 ozs. beets cooked until tender and crushed through a sieve. Keep the water in which they were cooked.

1 teaspoon chives cut in very small lengths.

1 teaspoon powdered dill.

10 prawns, can be replaced by 16 large shrimps.

1 teaspoon salt, 1/2 teaspoon pepper.

1 cucumber peeled, seeds removed and very thinly sliced.

2 cups sour heavy cream.

6 hard-boiled eggs sliced.

Add the cucumber to the beets and the water in which they were cooked, then the veal and its juice. Stir in the sour cream gradually, add the dill, salt and pepper, the chives, the prawns or the shrimps. Add the eggs carefully. Serve ice-cold.

CACIK (Turkish)

6 cucumbers peeled, seeds removed and cut in slices.

6 cups heavy yoghourt.

1 teaspoon salt.

6 tablespoons oil of sesame, a bland oil may be substituted.

Mix thoroughly and serve ice-cold.

TARATA (Greek)

3 green peppers, skinned and seeds removed.

6 egg plants, skinned and seeds removed.

Cook gently in 6 tablespoons olive oil without browning. Mash peppers and egg plants fine and mix thoroughly with 4 cups heavy yoghourt. Add 1 teaspoon salt, 1/2 teaspoon pepper, a pinch of cayenne, a pinch of powdered mint and 2 pressed cloves of garlic. Serve ice-cold.

After this chapter was completed further news of *gazpacho* came from Santiago de Chile in South America. Did the *conquistadores* take the recipe, along with their horses, to the New World? Señora Marta Brunet, a distinguished Chilean writer, is of Spanish or rather Catalan descent and she describes *gazpacho* as a meal of the Spanish muleteers. And meal it seems, in this version, rather than soup. These muleteers, she says, carry with them on their journeyings a flat earthenware dish—and garlic, olive oil, tomatoes and cucumbers, also dry bread which they crumble. Between two stones by the wayside they grind the garlic with a little salt and then add the oil. This mixture is rubbed all round the inside of the earthenware vessel. Then they slice the tomatoes and cucumbers and put alternating layers of each in the dish, interspersing the layers with layers of breadcrumbs and topping off the four tiers with more breadcrumbs and more oil. This done and prepared, they take a wet cloth, wrap it round the dish and leave it in a sunny place. The evaporation cooks the contents and when the cloth is dry the meal is ready. Too simple, my dear Watson.

VI.

FOOD TO WHICH AUNT PAULINE AND LADY GODIVA LED US

When in 1916 Gertrude Stein commenced driving Aunt Pauline for the American Fund for French Wounded, she was a responsible if not an experienced driver. She knew how to do everything but go in reverse. She said she would be like the French Army, never have to do such a thing. Delivering to hospitals in Paris and the suburbs offered no difficulties, for there was practically no civilian traffic. One day we were asked to make a delivery to a military hospital in Montereau, where we would lunch after the visit to the hospital. It was late by the time that had been accomplished and the court of the inn that had been recommended was crowded with military cars. When Gertrude Stein proposed leaving Aunt Pauline, for so our delivery truck had been baptised—not in champagne, only in white wine—in the entrance of the court, I protested. It was barring the exit. We can't leave it in the road, she said. That would be too tempting. The large dining-room was filled with officers. The lunch, for wartime, was good. We were waiting for coffee when an officer came to our table and, saluting, said, The truck with a Red Cross in the entrance to the court belongs to you. Oh yes, we proudly said in unison. It is unfortunately barring the exit, he said, so that none of the cars in the court can get out. I am afraid I must ask you to back it out. Oh that, cried Gertrude Stein, I can not do, as if it were an unpardonable sin he were asking her to commit. Perhaps, he continued, if you come with me we might together be able to do it. Which they did. But Gertrude Stein was not yet convinced that she would have to learn to go in reverse.

If Aunt Pauline had led us to Montereau on her first adventure, she was soon doing better. The committee of the American Fund had asked us to open a depot for distributing to several departments with headquarters at Perpignan. Aunt Pauline—Model T, bless her—made no more than thirty miles an hour, so we were always late at inns, hotels and restaurants for meals. But at Saulieu they would serve us for lunch *Panade Veloutée*, Ham Croquettes and *Peches Flambées*. They were cooked with delicacy and distinction. I got the recipe for

FLAMING PEACHES

Fresh peaches are preferable, though canned ones can be substituted. If fresh, take 6 and cover with boiling water for a few minutes and peel. Poach in 1 1/2 cups water over low flame for 3 or 4 minutes. Place in a chafing dish, add 1/4 cup sugar and 3/4 cup peach brandy. Bring to the table and light the chafing dish. When the syrup is about to boil light and ladle it over the peaches. Serve each peach lighted. This is a simple, tasty and effective dessert.

As we came into the dining-room I had noticed a man wandering about whose appearance disturbed me, he looked suspiciously like a German. German officer prisoners did occasionally escape. When the *maître d'hôtel* received our compliments for the fine cooking, I asked him who the man was and he said he was the proprietor of the hotel and had just been released from Germany where he had been a civilian prisoner. He had been *chef* for a number of years to the Kaiser, which not only accounted for the quality of the food but for the clothes which had misled me.

Aunt Pauline took several days to get us to Lyon where we were to lunch at La Mère Fillioux's famous restaurant. As a centre of gastronomy it was famous for a number of dishes, so La Mère Fillioux's menu was typical of Lyon. It is the habit in Lyon and thereabouts for restaurants and hotels to have set menus called *le diner fin* and *le dejeuner fin*, the choicest dinner and the choicest lunch. We had her choicest lunch, *Lavarets au beurre*,* hearts of artichokes with truffled foie gras, steamed capon with *quenelles* and a *tarte Louise*. Lyon is an

* *Note.* Lavarets are fish found in the lakes of Switzerland and the *Haute Savoie*.

excellent marketing centre. Fish served at lunch is caught in the morning, vegetables and fruits are of that morning's picking, which is of first importance in their preparation. Mère Fillioux was a short compact woman in a starched enveloping apron with a short, narrow but formidable knife which she brandished as she moved from table to table carving each chicken. That was her pleasure and her privilege which she never relinquished to another. She was an expert carver. She placed a fork in the chicken once and for all. Neither she nor the plate moved, the legs and the wings fell, the two breasts, in less than a matter of minutes, and she was gone. After the war, she carved a fair-sized turkey for eight of us with the same technique and with as little effort. When the fish appeared at our table she came to it and passed her hand about an inch above our plates to see that they were of the right temperature. Later she returned and with her little knife carved the largest and whitest chicken I ever saw. A whole chicken was always dedicated to each table, even if there was only one person at it. Not to have any small economies gave style to the restaurant. What remained of the chicken no doubt became the base of the forcemeat for the *quenelles* that were made freshly each morning.

STEAMED CHICKEN MÈRE FILLIOUX

The very best quality of chicken was used for steaming, as we use the best steel for gadgets, which is a very smart thing to do. The chicken has very thin slices of truffles slipped with a sharp knife between the skin and the flesh, and before trussing it the cavity is filled with truffles. Place the bird in the steamer over half white wine and half veal broth with salt and pepper and the juice of a lemon. The latter will give a flavour, but above all will keep the chicken white. The chicken was gigantic but so young that less than an hour had sufficed to cook it. This she told me when she came to carve it. She looked at it critically, then proudly. She was an artist.

Aunt Pauline eventually got us to Perpignan where we settled down to work. At the quiet hotel we had selected there was a banquet hall, closed for the duration of the war. I made arrangements to use it as a depot from which to distribute, the greater part to serve as a warehouse for the material already awaiting us at the station, and a corner to be screened off to serve us as an office where we could receive doctors and nurses who would come with lists of their individual needs. The hotel was delightful. There were wartime restrictions, and a few privations, but each guest was hoveringly cared for by one or more members of a family of four. The cooking was excellent, southern—not Provençal but Catalan. The Rousillon had been French for little more than 150 years. One of the local dishes was a dessert frequently served called

MILLASON

Pour slowly 2 cups boiling milk over 1 1/2 cups white cornflour and 1 cup sugar. Stir carefully to prevent lumps from forming. The mixture must be quite smooth. Boil stirring constantly about 20 minutes until quite stiff. Turn into a bowl, add 2 well-beaten eggs, 4 tablespoons melted butter and 1 tablespoon orange-flower water. Pour on to buttered platter and when cool enough to handle form into cakes and fry in oil in frying pan until golden on each side. Sprinkle with powdered sugar and serve at once.

They made the *millason* very large at Perpignan. They would be daintier if smaller. They are, surprisingly, not unlike our Southern fried corn bread.

The little lobsters in Perpignan were common, cheap and tender. They were cooked with a thick rich sauce and one day the very young waiter about to be mobilised was so eager to please that in rushing to serve us he all but spilled the sauce over my new uniform, of which I was inordinately proud.

PERPIGNAN LOBSTER

Cook 4 small lobsters not weighing more than 1 lb. each in boiling water, salted, for 18 or 20 minutes. During this time melt in a saucepan 4 tablespoons butter and heat in it 1 large carrot cut in thin rings and 2 medium-sized onions with a clove stuck on one of them and the white of 1 leek. When they are coated in butter sprinkle 1 tablespoon flour over them, mix well. Add little by little 1 cup hot dry white wine and 1 cup hot *bouillon*, 1 large bouquet of parsley, fennel and basil, salt, pepper, a pinch of cayenne, a pinch of saffron, 4 cloves of crushed garlic and 4 tablespoons tomato *purée.* Cover and cook slowly for 1 hour. Cut the lobsters longitudinally, take out the meat and place the 8 pieces in a hot casserole, take out the meat from the claws and place in the interstices of the lobster meat in the casserole. Take out of the sauce parsley, fennel and basil if you wish. They did not in Perpignan. Pour the sauce over the lobster meat into the casserole. Serve piping hot.*

There had been difficulty in getting gasoline on the coupons the army gave us. The major who was in charge of this distribution had been very helpful. Gertrude Stein did not like going to offices—she said they, army or civilian, were obnoxious. To replace her, I had introduced myself with her official papers and had allowed the major to call me Miss Stein. What difference could it make to him. We were just two Americans working for French wounded. By the time the difficulties had been overcome we had become quite friendly. At the last visit he said, Miss Stein, my wife and I want to know if you both want to dine with us some evening. It was time to acknowledge who I was. He drew back in his chair and with a violence that alarmed me said, Madame, there is something sinister in this affair. My

* *Note. Bouillon* is a "boiling," a stock made of veal, chicken or beef bones simmered in water with the special vegetables and herbs appropriate to the dish.

explanation did not completely reassure him, but Gertrude Stein waiting in Aunt Pauline in the street below would. I asked him if he wouldn't go down with me to meet her. He did. Her cheerful innocence was convincing, and his invitation was repeated and accepted. They were delightful. Madame de B. was training a local cook to cook as she believed cooking should be done.

During wars, no game is allowed to be shot in France except boar that come down into the fields and do great damage. To prevent this a permit is given to landowners to shoot them on their property. A farmer had shot one and brought the saddle to Madame de B. So we had

ROASTED SADDLE OF YOUNG BOAR

Even young boar is put in a marinade. One carrot and 1 onion cut in rings, 2 shallots cut in half, salt, pepper, a very large bunch of rosemary with just enough good dry red wine to cover the saddle is all that is needed for a light marinade. Four hours in the marinade turning the meat twice will suffice. An hour before it is time to put in the oven, take the saddle from the marinade and dry thoroughly. Strain the marinade into a saucepan, discard the vegetables but retain the branch of rosemary. For roasting allow 18 minutes a lb. in a 450° oven for the first 20 minutes, reduce to 350°. Roasting meat in an earthenware dish that can be brought to the table is a time-saver. Put a piece of butter in the dish that when melted will amply cover the dish. When the butter bubbles, place the saddle in the dish with 3 tablespoons melted butter on the meat. After 20 minutes baste with the branch of rosemary. If there is not enough melted butter and juice in the dish add 4 or 5 tablespoons hot marinade. Baste every 12 minutes, adding hot marinade as needed. While the roast is in the oven peel, boil and skin enough chestnuts to garnish the roast with a double wreath of them. Skim the juice before adding the chestnuts. Allow the chestnuts to cook in the skimmed juice for 15 minutes. Then serve in a sauce boat at the same time this

GAME SAUCE

Melt 1 1/2 tablespoons butter in a saucepan until it is golden brown, add 1 tablespoon flour, stir until light brown. Add slowly over low heat 1 cup of the hot marinade, the juice of 1 lemon, 1 tablespoon grated lemon peel, 1 tablespoon grated orange peel, a good pinch of cayenne, 3/4 cup currant jelly.

Venison may be cooked in the same way and pork is particularly good in this manner, except that the marinade should be of good dry white wine and the meat remain in it for 24 hours, turning four or five times. The game sauce is required for boar or venison, but should not be served with pork.

In one of the dark narrow back streets of Perpignan there was a small, remarkably good restaurant whose reputation was well known in Paris. After an excellent lunch we decided to ask Madame de B. and the Major to dine with us. Consulting with the *chef*, this was the menu decided upon:

Creamed chicken soup à la Reine Margot
Spring duckling with
Asparagus with Virgin Sauce
Coupe Dino

The *chef* generously gave me the recipe for Virgin Sauce which accompanied the steamed green asparagus.

VIRGIN SAUCE

For 1 person, place 5 tablespoons butter in a hot bowl, add 1/4 teaspoon salt, beat with a whisk until the butter foams, put it over hot but not boiling water for an instant. The butter must not melt. When the butter foams, add drop by drop, never ceasing to whisk 1 teaspoon lemon juice and 1 tablespoon tepid water. When they are well amalgamated with the foaming butter, add 1 tablespoon whipped cream and serve at once. This sauce is delicious with cold fish. It is something apart.

We had visited all the hospitals in the region and had reported on their future needs. Having made our last distributions we closed the depot at Perpignan and returned to Paris for another assignment. By this time, 1917, the United States had broken relations with Germany and had declared war. At last we were no longer neutral. On the road to Nevers, as Gertrude Stein was changing spark plugs—and when was one not in those days—we were told that a detachment of Marines was expected there that afternoon. Aunt Pauline was pushed to her utmost speed that we might be there for the entry. Thrilled by the first sight of the doughboys, we were unprepared for their youth, vigour and gaiety compared to the fatigue and exhaustion of the French soldiers. We were asked by some of their officers to meet the soldiers that evening and tell them about France. They had dozens of questions to ask, but what they wanted most to know was how many miles they were from the front and why the French trucks made such a noise. Though they were disappointed in our answers we had a wonderful and exciting evening together. It was their first contact with France and ours with our army.

In Paris the A.F.F.W. proposed we should open a depot at Nîmes where in advance of our arrival they would send several car-loads of material. News of our household was not so encouraging. During our absence our competent faithful Jeanne had gotten herself married. An excellent cook who worked by the hour consented to spend with us the few days we were to be in Paris. Severe rationing of meat, butter, eggs, gas and electricity had gone into effect. A small reserve of coal and assorted candles gave meagre heat and light. Ernestine accomplished much with little which permitted us to ask for lunch some of the Field Service men and volunteer nurses on leave in Paris. For them she made

KNEPPES

Remove skin and nerves from 1 lb. calves' liver, chop very fine, pound into paste, add 3 crushed shallots and 1 clove of garlic previously heated in butter, salt, pepper, 3 tablespoons flour, a pinch of mace, and mix thoroughly. Add one at a time the yolks of 2 eggs. When well amalgamated add gently the whites of 2 eggs well beaten but not dry. Drop by tablespoons into boiling salted water, and boil for 1/2 hour. They will rise to the surface. Drain thoroughly. Place on serving dish and pour over them 1/2 cup browned butter and 2 tablespoons dry breadcrumbs.

Ernestine said she learned this dish from a Belgian cook but we suspected he was of Alsatian origin. She also made for us

SWEETBREADS À LA NAPOLITAINE

Soak a pair of sweetbreads in cold water for an hour. Rinse and boil for 10 minutes in salted water. Rinse, remove all skin, cut into thin slices and brown lightly in 4 tablespoons butter in a saucepan. Add 1 cup sherry, 1 cup *bouillon*, 1 tablespoon tomato *purée*, salt, pepper, 1/2 cup diced ham. Place slices of sweetbread in this sauce. Cover and cook over low flame for 20 minutes. Prepare 4 thin slices Bologna sausage chopped fine, 1 large chopped onion, brown together in 2 tablespoons butter in an iron pot, add 1/2 lb. well-washed rice, turn with wooden spoon until rice is covered with butter. Add 1 cup boiling *bouillon.* Continue to stir until *bouillon* is absorbed, then add 1/4 lb. finely chopped mushrooms and 1 tablespoon *purée* of tomatoes, add 1 cup boiling *bouillon*, salt, pepper, a pinch of saffron. Continue to turn with fork slowly adding 2 more cups hot *bouillon*, 1 quart in all. When the rice has absorbed all the *bouillon*, it should be sufficiently cooked. Add 1/2 cup grated parmesan cheese and serve with sweetbreads.

The luxury hotel at Nîmes was in a sad way. The proprietor had been killed at the war, the *chef* was mobilised, the food was poor and monotonous. Aunt Pauline had been militarised and so could be requisitioned for any use connected with the wounded. Gertrude Stein evacuated the wounded who came into Nîmes on the ambulance trains. Material from our unit organised and supplied a small first-aid operating room. The Red Cross nuns in the best French manner served in large bowls to the wounded piping

HOT CHOCOLATE

3 ozs. melted chocolate to 1 quart hot milk. Bring to a boil and simmer for 1/2 hour. Then beat for 5 minutes. The nuns made huge quantities in copper cauldrons, so that the whisk they used was huge and heavy. We all took turns in beating.

Monsieur le Préfet and his wife, *Madame la Préfete*, whom we got to know and to like a lot, sent us word that a regiment of American soldiers was expected, that a camp was being prepared for them and that he would like us to be at the station with him when they arrived. Nîmes was agog with excitement and welcomed them as best it could—green wreaths, bunting and flags. Thanksgiving Day was some ten days after the soldiers arrived. Even the most modest homes were inviting our soldiers to lunch or to dinner to celebrate the day. That evening we had for dinner a large tableful of soldiers from camp. The manageress of the hotel, a large buxom blonde, had a group of American officers at her table. They were perhaps too noticeably gay.

At dinner one night—the inevitable whiting with its tail in its mouth was our monotonous fare—what appeared undoubtedly to be a German passed our table. This is really going too far, I said to Gertrude Stein. How dare an escaped prisoner show himself so publicly, so brazenly. Not your affair, let the authorities deal with him, she answered. After dinner the too-gay manageress said to me, There is a gentleman who has been asking to speak to you. I will send for him. It was the German. In perfect English he said he wished to speak to us alone a moment, and he pointed to some chairs. Gertrude Stein, always cheerful, agreeable and curious, sat down but not I. Who are you and what do you want of us, I asked. I do want some information from you, but first let me introduce myself. I am Samuel Barlow and we have several friends in common, but I am here as an officer in the secret service, in civvies naturally, to find out what is going on between yonder gay blonde and the American officers. The *Préfet* reported the case to us. He says he has reason to believe she is a German. Well, said I relieved, rather she than you. I mistook you for a German. My only civilian clothes, he said, were from Germany where I was a prisoner. This ended my concern with escaping German prisoners.

At Christmas the English wife of a prominent Nîmois organised, with the aid of the English companions and governesses who had posts in Nîmes, a dinner and dance for the British convalescent officers and men stationed there, and requisitioning for their army at Arles. After dinner we took turns dancing with the men. It was as gay as we could make it but the British Army was not cheerful. A few days later I had a visit from the prettiest of the young governesses. She said there had been an unfortunate incident after the party was over. She was preparing to turn out the light in her bedroom when there was a tap on the door which evidently connected with another room, and a voice asked, I say, Miss L., should I light my fire. Too surprised to answer, she was silent for a moment. The question was repeated, I say, Miss L., should I light my fire. Not for me, thank you, she answered. Of course the voice was unrecognisable, she ended, so I will never know which one of them it was.

Suddenly one day there was the Armistice and a telegram from the *Comité*—If you speak German, close the depot immediately, return to proceed Alsace civilian relief. If we had missed the spontaneous outburst of joy in Paris on Armistice Day we were going into liberated Alsace. One starlit morning we started in Auntie to make the six

hundred odd kilometres to Paris before night. Gertrude Stein ate her share of bread and butter and roast chicken while driving. Paris was still celebrating, and here the streets were commencing to be filled with the French Army, on the move into occupied Germany, and a certain number of Allied officers and men.

Having secured a German-French dictionary and fur-lined aviator's jackets and gloves, cumbersome but warm, we got off on the road again. The French Army was moving in the same direction Auntie was taking us. Near Tulle the mules dragging the regimental kitchen became unruly, swerved to the left and bumped into Auntie. A mudguard, the tool box and its contents scattered on the road and into the ditch. There was, of course, no way of recovering them. Starting off again, Gertrude Stein found the triangle so damaged as to make driving on the icy road not only difficult but possibly dangerous. We got to Nancy exhausted, too late for dinner, but Dorothy Wilde sweetly found two hard-boiled ducks' eggs, a novel but very satisfying repast. While Auntie was being repaired next day at the garage of the *Comité*, we had our first meal without restrictions. For a first course we were served

QUECHE DE NANCY

Prepare the evening before baking a pie crust made with 1 cup and 2 tablespoons flour, 5 tablespoons butter, a pinch of salt, 4 tablespoons water. The butter should be worked into the flour lightly but the mixture needs be no finer than rice. Roll on a slightly floured board into a ball, cover with waxed paper and put aside for at least 12 hours. Then roll lightly and fit into deep pie plate 9 inches in diameter. Place on the crust 1/2 cup cubed ham. Remove the skin from lean salt pork, cube 5 ozs., and place on crust having previously cooked the cubes in boiling water for 10 minutes, drained them and wiped them dry. Beat 3 eggs, pepper and salt, with 1/2 cup cream. Pour over ham and salt pork in pie crust. Dot with 12 small pieces of butter. Bake in preheated 450° oven for 15 minutes, reduce heat to 300° and bake for 20 minutes more. Remove from oven but not from pie plate for 10 minutes.

After which Aunt Pauline took us through no-man's-land to Strasbourg, still celebrating the Liberation. That night there was a torchlight procession of soldiers and civilians, the young girls in their attractive costumes (the black ribbon head-dress they had worn since 1870 changed to all the colours of the rainbow), with military bands. It was more like a dream than a reality. We were now in the land of plenty.

The next morning the director of civilian relief sent us to Mulhouse, the centre of the devastated area. Our material was already waiting there, and we got to work at once. For several days we unpacked the material, saw mayors, clergymen and priests from the ruined villages to which the refugees were returning on foot, by trucks, by any means they could find. It was very sad. Their determination and courage, however, were very heartening. We settled down to a winter of outdoor distributing to each village in turn. At Mulhouse we were not uncomfortable, first in the large hotel, then when that was requisitioned for officers in a purely Alsatian inn. There was an abundance of food, real coffee, large hams, real milk. Queues formed to look at them and buy them. The *pâtisseries* were filled with specialities of Alsace and the classic cakes of France. The French soldiers ate unlimited quantities and even sent them to their families and friends in France. At our inn they made a most satisfactory

SOUP OF SHALLOTS AND CHEESE

For each person lightly brown in butter on each side 1 slice of bread. Put in soup tureen, sprinkle with 1 tablespoon grated cheese and keep hot. Cook over low flame 4 sliced shallots in 1 tablespoon butter, add 1 teaspoon flour. Stir with wooden spoon, add 1 1/2 cups hot *bouillon*, cook covered over lowest flame, add salt and pepper, for 1/2 hour. Strain, add 2 tablespoons cream. Pour over bread and cheese in tureen and serve hot.

The quality of the material was excellent but there was no variety in the vegetables. They were all of the cabbage family, sauerkraut, cabbage, brussels sprouts and cauliflower. There were potatoes, to be sure, and apple sauce, which was considered a vegetable.

At Cernay we were helped in our distribution by little *Abbé* Hick, who had returned after the Armistice to find his church bombed, and the presbytery with the exception of one room in ruins. He asked us however to lunch with him the next time we were distributing in his neighbourhood. He met us at the door of his room and said, Welcome, come into the salon and warm yourselves. Excuse me while I go into the bedroom and wash my hands. He went to the far end of the room past a set dining-room table. Presently he returned and said, Now we will go into the dining-room and have lunch. All this without the least suspicion of the ludicrous. A refugee had cooked the simple but succulent lunch. The *abbé's* mother had sent him some good white wine from Riquewehr where she lived.

On Sundays we frequently lunched with the hospitable Mulhouseens who were gradually returning to the lives they had led before the war. Everything was in the French manner, with great elegance and luxury. They had really kept the manner of living of pre-1870. They had refused everything German. It was the memory of the way our French friends in San Francisco had lived come to life again.

At Monsieur B.'s there was for dessert, to my delight, a

TARTE CHAMBORD

Beat until foamy and thick with a rotary beater 1 cup and 1 tablespoon sugar and 8 eggs, gently stir in 2 cups and 3 tablespoons thrice-sifted flour. Add 1 cup and 1 tablespoon melted unsalted butter. Bake in a deep buttered and floured cake pan in 350° preheated oven for 30 minutes. Take from oven, let stand for

10 minutes, take out of pan, place on grill. When cold, cut horizontally four times, making five layers.

CREAM FOR CAKE

Turn with a wooden spoon in an enamelled saucepan the yolks of 10 eggs and very slowly add 1 1/2 cups icing sugar. Turn until thick and pale yellow. Put over lowest flame with 4 tablespoons butter for 2 minutes and as soon as butter is melted, stirring constantly, remove from flame and when the mixture is cold add drop by drop 3 tablespoons cold water. Return to flame stirring until the mixture is even. Remove from flame and when the cream is cold add drop by drop 1/4 cup kirsch or curaçao. Cover the layers and re-form the cake. Cover the top and sides with the cream and put in the refrigerator.

COFFEE FROSTING

In an enamelled saucepan put 3/4 cup very strong black coffee. Add enough icing sugar to make a very heavy cream. Warm over low heat. Pour on cake and with a spatula cover top and sides of cake. Sprinkle thickly with finely ground pistachio nuts.

We worked very hard all that cold winter distributing in the open air. Then one day there were fruit blossoms and storks. By this time the refugee relief was organised by the Government. We closed the depot, said goodbye to the officials and the people we had met and started off for Metz to see the battlefields of 1870 and to see Verdun. It was still a shambles. We wandered about locating the spots where the defence of the *poilus* had made history. It was the middle of the afternoon when Gertrude Stein finally asked, Where did you say we were going to lunch. I've gotten hungry. We got into Aunt Pauline and made our way slowly over the fields to something that had been a road. There we came upon a military car filled with officers. They said if we followed them we could find something to eat—in fact, they were eating there. They stopped at a corrugated iron hut and sure enough the man who presumably lived there made us an omelette with fried potatoes and a cup of real coffee, so rare in those days that at once I realised that the officers must have brought their own provisions with them and that we were sharing them. And then I remembered the two boxes of cakes the *abbé's* mother had sent to us the day before. So we got them out of Auntie. The little Alsatian cakes were of her own baking and delicious. We took a few of each kind and gave the rest to the officers whose unwitting guests we had been. These are their recipes.

SCHANKELS OR SCHENKELS

Cream 1/2 cup butter, add slowly 1/2 cup sugar, add slowly 4 eggs, one at a time. Add about 5 cups flour, depending upon size of eggs, with 1 teaspoon baking powder and 1 cup skinned and very finely ground almonds. The dough should be just firm enough to hold its shape when rolled in the hands into finger-length sausages. Fry in deep lard only enough of the cakes to cover the surface. Turn once to brown on both sides. Take from fat and place on absorbent paper. Sprinkle while still hot with plenty of icing sugar. They are a nice accompaniment to a glass of white wine or a cup of coffee. They keep well in a well-covered box.

LAEKERLIS

The Alsatians claim that Laekerlis are their creation, but the Swiss answer that they have two different kinds, one from Berne and one from Basle. This is Madame Hick's from Riquewehr. Warm 2 lbs. honey and skim, add 3 cups and 3 tablespoons sugar, 1 teaspoon cinnamon, 1/2 teaspoon cloves, 1/4 teaspoon powdered cardamom, 1/2 teaspoon allspice, 1 teaspoon mace, 1/4 teaspoon powdered anise, 1 cup finely chopped orange peel, 1/4 cup finely chopped lemon peel, 1/2 cup finely chopped citron and 2 cups finely chopped almonds. Mix thoroughly and gradually work in 7 cups sifted flour. Roll on a lightly floured board to 1/3-inch thickness. Cut in rectangles, place on lightly buttered baking sheet. If you have not a number of baking sheets, roll out the dough and leave on the floured board. Put aside for 24 hours in a temperate room and then bake in 350° oven. When baked, remove from baking sheet and place on grill. While still warm, paint with brush with this mixture:

Dissolve over very low flame 2 cups sugar and 3 tablespoons hot water. If this crystallises during the time the Laekerlis are baking, add a little hot water. These are of long conservation, as the French say.

We were lunching the next day with our friends from Nîmes, Madame T. and the *Préfet*, who were now installed at the *Préfecture* of Chalons-sur-Marne. We spent the night at the Hôtel Mère Dieu—a sacrilege in English. Chalons-sur-Marne is near Rheims and the wine cellar of the *Préfecture* is supplied by the Government with the best wines of the region. Lunch was served with ceremony and elegance worthy of the menu, the cooking and the wines. Of the menu I only remember the

SADDLE OF MUTTON MAINTENON

Put a saddle of mutton with salt, pepper and 3 tablespoons butter in a Dutch oven covered in a 350° oven. Turn every 10 minutes. Allow 10 minutes per lb. for the cooking of the saddle. When it is three-quarters cooked, remove from oven, place the meat on a carving board and with a very sharp knife slice very thinly both sides of the saddle. Be careful to lose none of the juice. Having cooked 1 chopped onion in butter, put it into 3/4 cup stiff *Béchamel* sauce with 1/2 cup chopped mushrooms cooked for 5 minutes in butter. Mix these ingredients thoroughly, spread on each slice of mutton, replace the slices on the saddle. Cover the saddle with three chopped onions, melted butter, breadcrumbs, more melted butter. Skim the juice in the Dutch oven, pour into preheated earthenware dish, place the saddle in the juice and the dish into a quick oven to brown the meat.

With this serve hearts of artichokes and small boiled potatoes *maître d'hôtel.*

As we were leaving the *Préfecture*, Gertrude Stein confided to me that she was going to show me a tank that the *Préfet* had told her was still in a field on the road to Rheims. It would not be much out of our way and it was certainly worth seeing. Auntie took fields so well. As we went along the national highway, Auntie and her driver were happily swaying and serpentining along. The wine at lunch was undoubtedly to blame for their lack of responsibility. They nevertheless negotiated the field. We saw the tank and got on our way to spend the night with Mildred Aldrich at the Hillcrest. It was from there she had seen the first Battle of the Marne and the German retreat. In her garden that evening I wrote the last report to the *Comité.*

The next morning we were back in Paris, more beautiful, vital and inextinguishable than ever. We commenced madly running about, to see our friends and theirs. It was gay, a little feverish but pleasurably exciting. Auntie Pauline took us to lunch and dinner parties. Our home was filled with people coming and going. We spoke of each other as the chauffeur and the cook. We had no servant. We had largely overdrawn at our banks to supply the needs of soldiers and their families and now the day of reckoning had come. We would live like gypsies, go everywhere in left-over finery, with a *pot-au-feu* for the many friends we should be seeing. Paris was filled with Allies, the Armies, the Peace Commission and anyone who could get a passport. We lunched and dined with a great many of them, at their messes, headquarters, homes and restaurants. One evening Aunt Pauline had taken us out to the Bois de Boulogne to dine with friends in the garden of one of its restaurants. While dinner was being served the *maître d'hôtel* asked me to please follow him, someone wished to speak to me. It was a policeman to announce that trucks were not allowed in the Bois. They had been tolerated during the war, but an Armistice had been signed. So would Madame see that her truck did not appear there again. When I got back to the table an excellent dish was being served.

HARICOT

(yes, that is its seventeenth-century name)

Take an oxtail, separate the joints, put them in a pot with some marrow, salt, 1 clove, a twig of sage, 1 laurel leaf and a little water. When the meat is half cooked add 1 lb. sliced turnips, 1 lb. peeled and skinned chestnuts and 10 slices of any highly spiced sausage. Cook until the meat is tender and the juice reduced. Then add 8 slices of toast on which 3 tablespoons vinegar have been sprinkled. There are some who like a few prunes or raisins added. Our haricot's sauce had raisins in it. They had previously been swollen by soaking in hot water. They are an agreeable addition and cut the acid of the vinegar.

In the spring of 1919 we went to Normandy to stay with friends. A calf that they wished to sell at the local fair was put into Auntie, and she brought the first potato harvest to market, so that a thorough cleaning of Auntie was required on our return. On the way to Paris we stopped at Duclair. The hotel was on the Seine, its cooking was famous. It was there we had

SOLE DE LA MAISON

Poach gently in milk the fillets of sole with salt and pepper. Cover and simmer gently for about 15 minutes, depending upon the thickness of the fillets. Drain thoroughly. Place on a preheated carving dish and keep hot. Poach only long enough to heat 4 oysters and 4 large shrimps for each fillet. Place them alternately on the fillet. Cover with heavy cream sauce made with heavy cream and flavoured with 2 tablespoons best dry sherry.

At Duclair everything was cooked in cream: chicken, cabbages, indeed all vegetables and most meats. We stayed there several days before this bored us. At nearby Rouen butter replaced the cream. The butter was of such an excellent quality that it seemed advisable to make arrangements to have a weekly delivery to us in Paris. At a recommended creamery this was discouraged. Parcel post was not yet reliable. So I bought 12 lbs. to take back to friends and for our use. To my delight each pound was packed in a porous black earthenware jar of exactly the same form as the Gallo Romaine ones that we had been seeing in the museums. I kept two of them until last year when I gave them to a friend who would not believe they were not originals and that I was not parting with treasures.

When we returned to Paris our friends convinced us that the time had come to transform Auntie into something more suitable for the use we were making of her. The state of her engine did not warrant the purchase of a new body. We had her high canvas cover lowered so satisfactorily that we had her painted, but, neglecting to choose a colour, she returned to us painted a funereal black. This and her new form suggested a hearse—for an *enterrement de troisième classe.* She would continue to be *risible* to the end.

Gradually we realised that poor Auntie was weakening. It was no longer advisable to take her too far on the road. We would go to Mildred Aldrich's and in summer have a picnic lunch in her garden, and indoors in her cosy little home in winter when Amélie, her devoted friend, neighbour and servant, would make us the very best we ever ate

CRÈME RENVERSÉE

Put 4 lumps of sugar in a metal pudding mould over a very low heat. When melted add 1 1/2 teaspoons cold water. Turn the mould in all directions to cover it completely with the caramel. Heat 2 cups evaporated milk with 1/2 cup sugar. Put aside to cool. Stir 4 eggs until thoroughly mixed, add 2 teaspoons orange-flower water. Strain into the cold cream. Pour into prepared mould, set mould into pan of hot water reaching to half the height of the mould. Place in preheated 350° oven for 40 minutes. The water should not boil. When a knife gently stuck into the *crème* comes out dry, remove from oven and remove mould from water. Do not attempt to turn out of mould until cold. This is very nice served with chocolate sauce.

I was aghast to find Amélie using tinned milk in a country of excellent fresh milk. She explained that it was the milk Madame's friends had sent her in such quantities after the war. She assured me that even fresh milk did not adequately replace it. It was the only time in my experience that a French woman recommended American tinned products to replace French fresh ones.

Auntie held out for another year and then one day as we were passing the entrance to the Palais du Luxembourg she stopped short. Nothing Gertrude Stein did was of any avail, she would not budge. We were quickly surrounded by an amused crowd and by half a dozen not at all amused policemen. One of them asked if we didn't know that it was an infraction of the law to obstruct the entrance to a public building, particularly the Senate, where the Prime Minister was expected any minute to drive through. Indeed soldiers on motor bicycles had already arrived and a large car was being held back. We jumped out of Auntie, the police shoved her out of the way and the big car passed through with Monsieur Poincaré's beautiful head out of the window to see the cause of the commotion. We basely deserted

Auntie and went home on foot. Gertrude Stein telephoned to the garage to haul her in and repair her at once. The answer next day was that she was beyond repair. Nothing daunted, Gertrude Stein went to the garage with our good friend, Georges Maratier. She wanted his help and his advice to realise it. He said the two of them would drive to the country garage of his parents where there was plenty of room. There Auntie would be an honoured war souvenir. Her odyssey was the subject of the following winter's conversation. She is still there, but I have never had the courage to go to see her.

At no matter what sacrifice it was unthinkable that we should be without a car. Fords were still scarce in France, but Gertrude Stein inveigled a promise that she would have a two-seater open one within two weeks. As we were driving her to a beautiful new box fortunately secured in our neighbourhood for her I remarked that she was nude. There was nothing on her dashboard, neither clock nor ashbox, nor cigarette lighter. Godiva, was Gertrude Stein's answer. The new car was baptised without benefit of clergy or even a glass of wine. The reason for her name soon disappeared with all the gifts she received, but Godiva remained her name.

Now we would go on excursions out of town again. On the road to Chartres we made acquaintance with an excellent little restaurant which unfortunately disappeared during the Occupation of the second war. There we ate

CHICKEN SAUTÉ À LA FORESTIÈRE

Put the chicken with 3 tablespoons butter in a Dutch oven over medium heat, keep turning it about. When lightly browned on all sides reduce heat and cover, in 1/4 hour add 1/2 lb. morels previously cooked in tepid water, brushed and well rinsed, 1/3 lb. pig's fat cooked previously for 5 minutes in boiling water and well drained, salt, pepper. Cover and simmer over low flame for 3/4 hour, depending upon size of chicken. When done remove from flame, place chicken on preheated serving dish, remove morels from sauce with perforated spoon and place around chicken. Skim juice and return pot to stove, add 1 cup good dry white wine and 1/2 cup stock. Boil uncovered for 5 minutes. Strain and pour over chicken. New potatoes browned in butter are almost obligatory for this dish.*

Godiva took us to Orleans where on the banks of the Loire we ate freshly caught

SALMON WITH SAUCE HOLLANDAISE AU BEURRE NOISETTE

The salmon was cold, decorated with tomatoes, hard-boiled eggs (yolks and whites) pounded in a mortar separately and thinly sliced cucumbers.

THE SAUCE

Sauce Hollandaise is easily and quickly prepared if you pay careful attention to this foolproof recipe. Put 4 yolks of eggs and a little pepper and salt in a small saucepan over the lowest possible flame. Stir continuously with a wooden spoon, adding drop by drop 1/2 lb. browned butter. Put 3/4 cup shelled hazel nuts in the oven. When they are warm remove from oven and roll in a cloth until all the skins are removed. Pound them in a mortar to a powder, adding from time to time a few drops of water to prevent the nuts from exuding oil. Strain through hair sieve. Replace in mortar and add 1 cup water. Mix with pestle or wooden spoon. When perfectly amalgamated commence to add in very small quantities at a time to the egg yolks in the saucepan, stirring continuously. If the contents of pan become too hot remove a moment from flame and add a small quantity of butter to cool the mixture

* *Note.* Morels are edible fungi.

before replacing over flame. When all the butter has been incorporated remove from flame and slowly stir into the sauce 1 tablespoon vinegar. Serve in sauce boat.

This is a rare sauce. Once the hazel nuts are prepared it takes little time to prepare. Do not take the time to think that almonds can successfully replace the hazel nuts which give the sauce its elusive and distinctive flavour.

Even though Godiva was what a friend ironically called a gentleman's car, she took us into the woods and fields as Auntie had. We gathered the early wild flowers, violets at Versailles, daffodils at Fontainebleau, hyacinths (the bluebells of Scotland) in the forest of Saint Germain. For these excursions there were two picnic lunches I used to prepare.

FIRST PICNIC LUNCH

A chicken is simmered in white wine with salt and paprika. Ten minutes before the chicken is sufficiently cooked add 1/2 cup finely chopped mushrooms. When cooked remove chicken and drain. Strain mushrooms. The juice may be kept in the refrigerator to be used as stock. Put the mushrooms in a bowl, add an equal quantity of butter and work into a paste. This is very good as a sandwich spread or may be thoroughly mixed with the yolks of 3 hard-boiled eggs and put into the hard-boiled eggs which have been cut in half.

For dessert fill cream-puff shells with crushed sweetened strawberries.

SECOND PICNIC LUNCH

One cup finely chopped roast rare beef, 1 teaspoon chopped parsley, 1 teaspoon crushed shallots, salt, pepper, 1 teaspoon tomato *purée*, 1 tablespoon sour cream, a pinch of dry mustard. Mix thoroughly. Lightly toast on one side only eight slices of bread.

Butter generously the untoasted sides. Spread on the buttered side of four slices of the bread the meat mixture. Cover, with buttered side over meat, with the other four slices of bread.

To eat with these sandwiches, prepare lettuce leaves on which boiled diced sweetbreads are placed, 1 1/2 cups for four large lettuce leaves. On the sweetbreads place 4 chopped truffles that have been cooked in sherry. Roll the lettuce leaves round the sweetbreads and truffles, neatly trim with scissors and tie with white kitchen string in three places.

For dessert peel apples, core, cut in half and caramelise in 3/4 cup sugar and 1/4 cup water that has boiled to the caramel stage, for about 10 minutes. Completely coat the apples with the caramel. When dry wrap in square of puff paste, moistening the edges so that they will adhere. Fry in deep fat until golden brown on all sides. Remove from fat to absorbent paper. While still hot cover generously with sifted icing sugar. Excellent hot or cold.

Godiva had been taking us successfully to places in the neighbourhood of Paris. It was time to give her a wider field. In early spring she would take us to the Côte d'Azur. We had been asked to stay with a friend at Vence. We would wander down the Rhône Valley and see to what she would lead us. We would start early and spend the night at Saulieu to which Auntie had taken us several years before. To look at the church when we got there we parked Godiva in the square about 100 feet from the hotel. When we returned Godiva flatly refused to start. What were we to do. A red-liveried groom appeared and asked if he could help. Perhaps if he pushed the car—which he did. Godiva's engine started. Before we knew what she was up to we were in the court of what turned out to be the rival hotel of the one we intended to go to. It was her first display of instinct to lead us to the real right place. It must be acknowledged that never later did she shine with equal lustre.

The Côte d'Or then had as its proprietor and *chef* a quite fabulous person. First of all he looked like a great Clouet portrait, a museum piece. He had great experience and knowledge of the history of French cooking from the time of Clouet to the present. From him I learned a great deal. At dinner that evening we realised that he was one of the great French *chefs*. Each dish had a simplicity and a perfection. Comparing the cooking of a dish to the painting of a picture, it has always seemed to me that however much the cook or painter did to cover any weakness would not in the least avail. Such devices would only emphasize the weakness. There was no weak spot in the food prepared by the *chef* at the Hôtel de la Côte d'Or.

For dinner we had

MORVAN HAM WITH CREAM SAUCE

Four thick slices of ham from which the skin but not the fat have been removed are placed in a saucepan and browned lightly in butter with 1 onion, 1 carrot, 1 leek and the greens of 6 radishes. Add 1/2 cup Madeira or good dry sherry and 1 1/2 cups *bouillon*, salt, pepper and 1 crushed shallot. Cover and cook over very low flame for 2 hours. Be careful that it does not burn. If the pan is hermetically covered this will not happen. If not, it may be necessary to add more wine and *bouillon* in the above proportions. At the end of 2 hours remove from flame. Strain juice, reject vegetables, put ham aside. Skim juice and place 3/4 cup over medium heat uncovered. Reduce to 1/2 cup and place ham in saucepan. Glaze on both sides. Add rest of strained juice and 1 cup heavy cream. Bring to a boil and simmer for 2 minutes, tipping the saucepan in all directions.

Saulieu is in the Morvan, an old division of France and part of Burgundy and has always been famous for its ham, which is not unlike York ham.

We stayed on next day for lunch and again chose a simple dish

THREE-MINUTE VEAL STEAK

Ask the butcher to cut very thin slices in a fillet of veal, remove bones, skin and fat. It is well to count upon two slices per person, eight slices for four people. Brown on both sides in 1/4 cup butter in Dutch oven over high flame, salt and pepper. When they are brown, cover and put in preheated oven at 400° for 5 minutes. Add 1 cup hot dry white wine. Take meat from oven and place on preheated serving dish. Skim juice, place over high flame and mix well with glaze at the bottom and sides. Reduce heat to very low, add in small pieces 6 tablespoons butter, shaking pot in all directions. Add a squeeze of lemon juice and pour over meat and sprinkle 4 tablespoons chopped parsley over meat and sauce. This is delicious if the preparation is not allowed to drag.

At Mâcon that evening for dinner we had

PURÉE OF ARTICHOKE SOUP

Wash thoroughly 6 large artichokes, cut them in half vertically and remove chokes with a sharp knife. Put 3 tablespoons butter in a saucepan. When melted add artichokes. Stir them with wooden spoon until well covered with butter. Add 3 cups hot water and 3 cups hot chicken broth. Cover and boil steadily for 1 hour. Then add 2 cups thickly sliced potatoes. Cook for 1/2 hour more. Remove from fire and with a silver spoon scrape from all the leaves all the edible bits of artichoke. Crush this with the hearts and potatoes through a hair sieve with a potato masher. Strain juice in pan and add to strained artichokes and potatoes. Wash pan and place strained material in it. Heat over medium flame. If too thick add more chicken broth. Add salt and pepper. Before serving reduce heat and add 1 cup butter in small pieces. Tip saucepan in all directions and serve in preheated tureen in which you have placed very small unbuttered *croûtons.* This soup is well worth the effort and time it takes to make it.

The *chef* at Mâcon was proud of his desserts. They were delicious, varied and abundant. He would come to your table as one after another was presented and his feelings would be hurt if you did not at least taste each one of them. There were always chocolate, coffee (or Mocha), caramel and pistachio creams and ice creams, berries in season with heavy but unwhipped cream in which a spoon stood upright, *tartes* of all kinds and one cake—a *Gâteau de la Maison.* For years this cake was a puzzlement to me. It wasn't until Lord Berners brought us one to Bilignin one summer when he was going to stay with us that I had enough courage to attempt an approach to the famous cake. How he had inveigled that cake out of the Mâcon hotel was not explained, but one suspected. This is as near as my experiments got me to

THE MÂCON CAKE

Brush four shallow-layer cake tins lightly with melted butter. At once sprinkle with sifted flour. Rap on back of tins to remove excess flour. Mix 1 1/4 cups powdered almonds and 1 1/4 cups sugar. Put aside. Put the whites of 8 eggs in a bowl and commence to beat them, gradually increasing the speed. Do not stop beating for an instant. When done they should form a stiff peak when whisk is removed. With a wooden spatula lightly mix in the sugar and powdered almonds. Fill the four layer pans and put at once in a preheated 300° oven for about 1/2 hour.

BUTTER CREAM FOR THREE LAYERS

Boil 3/4 cup water with 1 cup sugar for 10 minutes. Stir yolks of 8 eggs for 5 minutes and slowly add syrup. Pour into double boiler stirring continuously with wooden spoon until spoon is coated. Strain through fine sieve beating vigorously until cool. Then add 1 cup whipped cream. Put 3 cups butter in a heated bowl and beat until creamy. Then very slowly add the syrup, yolks of eggs, whipped-cream mixture. Put aside.

Mocha Cream

Take a third of above mixture and drop by drop add 4 tablespoons very strong black coffee. Spread 1 1/2 cups of this evenly on one of the meringue layers. Put aside the remaining mocha cream. Cover with another layer.

Kirsch Cream

Take another third of butter cream and add drop by drop 2 tablespoons best kirsch. Spread a third of this on layer of meringue covering mocha cream. Put aside the rest of kirsch cream. Cover kirsch cream with a third meringue layer.

Pistachio Cream

Take the remaining third of butter cream and add 3/4 cup thrice-ground pistachio nuts. Spread a third of this evenly on layer of meringue covering the kirsch cream, and cover with fourth and final layer of meringue. Reserve the rest of the pistachio for the crowning operation which is

TO DECORATE THE CAKE

Take remaining mocha cream and spread evenly over a third of top of cake and a third of the sides of cake. Spread evenly remaining kirsch over centre third of cake and centre third of sides of cake. Spread evenly remaining pistachio cream over remaining uncovered part of cake and sides. Now form a design in centre of cake about 2 inches in diameter of crystallised apricots and angelica. It is effective to make petals of the apricots with surrounding leaves of angelica. On the outside of the cake make very small flowers of the apricots with surrounding small angelica leaves, one little bouquet for each slice of cake. Keep in a cool place until time to serve. The meringue layers can be baked in advance. This is of course not a cake but a dessert. It has an elusive subtle flavour and is quite worth the time it takes to make it.

At Mâcon we heard of a very small but highly recommended restaurant at Grignan, a village of six hundred inhabitants in the Drome, which is a department of France of fine cooking and romantic landscape. We wired to friends to meet us at Grignan for lunch. The name was familiar—was it not the name of Madame de Sevigné's adored daughter? In the guide book I found that the Château de Grignan was still intact, that one could visit it, and that Madame de Sevigné was buried in the church in the village. We would make a pilgrimage to the spot after lunch. When we arrived in the village and saw how small the restaurant was, we wondered if there would be room for the four of us. Madame Loubet, the proprietress and cook, was of commensurable size. Like many first-rate women cooks she had tired eyes and a wan smile. This seemed a happy omen. She said for lunch there would be an omelette with truffles, a *fricandeau* of veal with truffles, asparagus tips and a local cheese. The little restaurant was of the seventeenth century, the uncovered tables and chairs of the same period. We said it was Shakespearian. So did our friends when they arrived. We were enchanted with the *décor.* Lunch would be worthy of it.

MADAME LOUBET'S ASPARAGUS TIPS

Early spring is the time for the first small green asparagus, very like the wild ones. Wash quickly—do not allow to remain in water—discard white stems. Tie into neat bundles, plunge into boiling salt water. Allow about 8 minutes for their cooking. They should not be overcooked; much depends upon their freshness. Put aside. Over very low flame put in a saucepan 4 tablespoons butter (for 1 lb. asparagus). When butter is melted, add asparagus tips still tied in bundles. Add 4 tablespoons heavy cream. Do not stir, but gently dip saucepan in all directions until the asparagus are coated with butter and cream. Then remove from flame. Place asparagus on preheated round dish with the points facing to the edge of the dish. Gently cut the strings with kitchen scissors. In the centre place 1/2 cup heavy whipped cream with 1/2 teaspoon salt mixed in it. Serve before cream has time to melt. This is a gastronomic feast. And a thing of beauty.

The cheese called Cochat, a speciality of the region, is made of the milk of very young ewes, and ripened in vinegar. It is then pressed under weight and served in the shell of a medium-sized onion. With this it is traditional to drink a red wine, preferably a good vintage of Châteauneuf du Pape.

Madame Loubet's cooking was delicate and distinguished, and we often returned to enjoy it. We would find nothing comparable before the end of our journey. That evening at Marseilles a bowl of soup sufficed.

After a long run down the coast of the Mediterranean we arrived at Vence to find a numerous party for dinner. Our friend was something of a *gourmet*, his Belgian cook had a well-organized kitchen and produced varied and succulent menus.

The vegetable garden was already producing spring potatoes, string beans, artichokes, salads and, before we left, asparagus. The gardener amiably allowed me to gather each morning the day's vegetables. It takes a long time to gather enough very young string beans for eight or ten people. Between the vegetable garden and the rose garden my mornings were happily occupied. To me this pleasure is unequalled.

From Vence we frequently drove down to Nice to have a fish lunch at a small and unpretentious restaurant on the sea. For us they made a local dish—

GRILLED PERCH WITH FENNEL

Wash and completely dry a perch weighing about 3 lbs. Rub salt and pepper inside the fish. Paint it with melted butter, paint the grill with butter. Place fish on grill and grill under flame for 25 minutes turning the fish twice and painting it with butter each time. Have a quart of fennel greens washed and dried thoroughly. When the perch is cooked place on a preheated metal dish that withstands flames—not one of pewter, for example. Pour 1/2 cup melted butter over fish and completely cover with fennel greens. Take to the dining-room, light the fennel leaves. When flaming, serve. It is exciting and delicious, one of the rare Provençal dishes into whose preparation garlic does not enter.

On our trip back to Paris Godiva was no longer inspired. It was we who were obliged to take the initiative. As we were in haste we took no time to go out of the way to discover new places. We contented ourselves with the tried-and-not-found-wanting than which there is nothing more deadly. Once in Paris again she returned to her competent leadership. She took us to Les Andelys where we lunched out of doors at a *bistro* (café-restaurant where coarse and rarely good food is served) on fried fish caught from the Seine just below the terrace where we were lunching. Fish was followed by a really tender Châteaubriand and *soufflé* potatoes. Back in Godiva and on the road again it was obvious that somewhere we had made a wrong turning. Was Godiva or Gertrude Stein at fault? In the discussion that followed we came to no conclusion. We were on the road to Nogent-le-Rotrou. We would see what we would find there. It was an enchanting landscape of thatch-roofed villages, fields coloured with the first poppies and cornflowers and hedges of blossoming hawthorn. Nogent-le-Rotrou was old, clean and sympathetic. The hotel was furnished with pale Restoration furniture, small figurines and wax flowers under globes were in all the rooms. The food was simple but skillful. We stayed there several days. It was a woman who cooked, quietly and expertly. She showed me how to make

RILLETTES

Grind in meat chopper 2 lbs. breast of pork. Melt in iron pot 1 lb. lard. When a pale gold, add chopped pork, 1 tablespoon salt, 1 teaspoon pepper, 1 teaspoon powder for poultry dressing. Simmer uncovered over very low heat for 4 hours, stirring to prevent burning. After 4 hours remove from flame. When cold enough, ladle into jelly glasses. See that meat is evenly distributed. When completely cold, cover with paper. In a cool place the *rillettes* will keep several months. They are nice served with salad or as *hors d'oeuvre* or for sandwiches.

On leaving Nogent we took dust roads through the same landscape we had driven through to get there. We were on our way to Senonches, on the edge of a forest. We were seduced at once by the little town, the hotel and the forest. We not only ordered lunch but engaged rooms to spend the night. While waiting for lunch to be cooked, we walked in the forest where Gertrude Stein, who had a good nose for mushrooms, found quantities of them. The cook would be able to tell us if they were edible. Once more a woman was presiding in the kitchen. She smiled when she saw what Gertrude Stein brought for her inspection and pointed to a large basket of them on the kitchen table, but said she would use those Gertrude Stein had found for what she was preparing for our lunch,

A FLAN OF MUSHROOMS À LA CRÈME

Lightly mix 1 cup and 1 tablespoon flour, 4 tablespoons butter, 1/2 tablespoon salt and 1 egg. Gently knead this dough on a floured board and roll out. Add 1 tablespoon heavy cream, knead and roll out. For the third and last time add 1 tablespoon heavy cream, knead and roll out. Roll into a ball, lightly flour, put in a bowl, cover and put in a cool place for 1 hour. After an hour's repose, roll out and bake in deep pie plate in preheated oven at 400°. In the meantime prepare

SAUCE MORNAY

Put 2 tablespoons butter in a saucepan over low heat. When melted add 1 medium-sized onion cut in thin slices, 1 medium-sized carrot cut in thin slices and a stalk of celery cut in thin slices. Turn with a wooden spoon until the vegetables are lightly browned. Add salt and pepper and 1 1/2 tablespoons flour. Turn with a wooden spoon for 5 minutes. Then add slowly 3 cups hot milk. Be careful that there are no lumps. Simmer for 1/2 hour, stirring frequently to prevent burning. Remove from flame, strain and discard the vegetables. Add 3 tablespoons heavy cream to strained sauce and 1/2 cup grated Parmesan cheese. Wash and brush carefully but do not peel 1 lb. small mushrooms. Drain well and wipe dry. Melt 1 tablespoon butter in saucepan, add juice of 1/2 lemon and 1 tablespoon sherry, salt and a pinch of pepper, paprika, the mushrooms and 1/2 clove of mashed garlic. Cover and cook over low flame for 8 minutes. With a perforated spoon remove mushrooms and mix with Mornay sauce. Pour into baked pie crust and place in preheated 450° oven for 12 minutes. Be careful the bottom of the crust does not burn—an asbestos mat under the pie is a protection.

This is a dish that every experienced cook in France prepares in his own way. If morels (edible fungi) are used the cheese is omitted from the sauce. A tomato sauce may be substituted for the Mornay sauce but is not as fine. Chopped ham may be added to the sauce or 4 or 5 boned and crushed anchovies. The variations are endless. The crusts vary too from a biscuit dough to a puff paste. It is a dish that is always well received.

During the winter two of our friends, Janet Scudder the

sculptress and Camille Sigard of the Metropolitan Opera of its great days, suggested that they too buy a two-seater Ford and that in summer we should all go south. Janet was hunting a home, the house of her dreams. She was an admirable travelling companion and had a gift for locating good food and first-class wine. When summer came we started off one sunny day to lunch at a restaurant that Janet knew. It was indeed a good lunch but the view from the balcony where lunch was served was too distracting for the enjoyment of a well-prepared meal. We once stayed in the country with friends who had two young sons. At table they chattered to each other continuously but so quietly as not to interfere with our conversation. One day, sitting next to one of them who was silent, I asked why he was not talking to his brother. He explained that he never did when there were artichokes with *Sauce Mousseline.*

Janet in her search for a house would uncover a small *bistro*, but the food was not coarse at the ones Janet took us to. It became a joke whether she and not Godiva had the instinct for achieving the gastronomic bull's-eye. We were on dust roads all the way to Avignon. Maybe we helped, but Godiva led the way to Aramon, a village dominated by a fortified castle, where we lunched roughly but tastily on a dish of the region,

HEN À LA PROVENÇALE

Take a not too old hen and cut into joints, 1 lb. breast or shoulder of mutton, 3 tomatoes, 3 hearts of artichokes, 3 small vegetable marrows, 3 sweet peppers, 1 cup chick peas that have been previously soaked, 3 turnips, 3 medium-sized onions with a clove stuck in each one, 1 teaspoon Spanish red pepper, 1/4 teaspoon cumin powder, salt and a pinch of cayenne. Put in a pot, cover with water, place over medium flame covered. When it boils reduce to low flame and simmer for 2 hours. Serve in a deep dish with chicken and mutton in centre, the vegetables around. Pour as much of the juice as the dish will hold, the rest in a sauce boat. This is not only nourishing and succulent but sufficiently satisfying, with cheese and coffee to follow, to make a meal.

Our friends not yet having found a home were ready to try the Côte d'Azur, expecting us to go too. But neither Gertrude Stein nor I found the Mediterranean coast sympathetic. The part of France that had seduced us by its beauty lay between Avignon and Aix-en-Provence, Orange and the sea. Saint Rémy would be a point from which to radiate. Janet asked if we had ever eaten there. We were obliged to answer that we had not. We would find out. We had not selected it for its culinary possibilities. Saint Rémy and the country about had a poignant beauty that would compensate for the deficiencies of the inn. The rough bare rooms they showed us and which I bespoke at once looked out on a pleasant garden. Lunch was mediocre. Janet ominously remarked that we would regret our choice. In the afternoon our friends drove over to Aix-en-Provence and we settled down for a long stay. We were where we wanted to be.

The *mistral* blew and the food was bad, but we were enchanted with our walks and drives in all directions. We commenced to look for a passable restaurant or *bistro*. After investigating the provision shops we concluded it was not reasonable to suppose that it could be a country of good cooking. This did not discourage us, it was just a fact. Marseilles was within easy reach and we could run down there for diversion, shopping and a good lunch.

We had had no news from the friends at Aix-en-Provence except a telephone message asking how we were supporting life at Saint Rémy, when a telegram from Janet announced that we should come over at once to see the house she had found and was busy buying. We drove over next morning. The commonplace little house was built in a hollow—therefore without a view—in a large tract of treeless uncultivated land. Because it was unlike the taste of our friend we tried to dissuade her from buying it, but it had become a fixed idea. We were obliged at lunch to listen endlessly to Janet and her new home. Soon we were going over to see the interior decorations under way and what had been achieved in the well-planned garden.

During this time we had gone down to Marseilles and had tried at two of its best restaurants Marseilles' unique creation,

BOUILLABAISSE

The fish should be more than fresh, it should be caught and cooked the same day. This is what gives the dish its quality. There must be many different kinds of fish to give the proper flavour. It is not only the ingredients that go into the sauce—which is not a sauce but a soup—it is the flavour of the fish that predominates. There should be at least five different kinds of fish. In Marseilles where the *Bouillabaisse* was born there are frequently seven or more not counting the shellfish. It can not be repeated too often that they must be very fresh. In France there are three different kinds of *Bouillabaisse*—the unique and authentic one of Marseilles with Mediterranean fish, the one of Paris made of fish from the Atlantic, and a very false one indeed made of fresh-water fish.

Take 5 lbs. gurnards, red snapper, red fish, mullets, pike, turbot and dory, wash, scale, cut off the fins and heads. Cut the large fish in 1-inch slices, leave the smaller ones whole. Have two extra heads of any large fish. Wash very thoroughly and put them with the heads of the small fish with 1 carrot, 1 onion, 1 laurel leaf, a twig of thyme. Cover with 7 cups cold salted water, bring to a boil uncovered. Skim and cover, boil until reduced to half. Then mash with potato masher through fine sieve. Boil 1 large lobster and 1 crab. When cooked remove from water and drain. Cut the meat from the body of the lobster into four pieces. Put the meat from the crab and from all the claws together. Pour into a large saucepan 1/4 cup olive oil. When it is hot add 3/4 cup thinly sliced onions, 3 crushed shallots, 3 cloves of crushed garlic, 1/2 sweet pepper (seeds removed), 1 large peeled tomato cut in slices, 4 stalks of celery, 1 two-inch slice of fennel. Turn with a wooden spoon until well coated with oil, then add 1/4 cup olive oil, 3 twigs of thyme, 1 laurel leaf, 2 whole cloves, 1 piece of the zest of an orange, salt and pepper. Over very high heat add the *bouillon* of the fish heads. Boil covered for 5 minutes. Put the less quickly cooked fish into the saucepan. Boil furiously uncovered for 5 minutes. Add the rest of the fish and the lobster meat from the body. Boil furiously uncovered for 5 minutes. Remove from heat. Remove fish and lobster with perforated spoon, wipe whatever may be adhering to them. Place in a large deep dish and keep hot. Strain juice from saucepan, replace on stove, add crab meat and meat from clams. Put in a bowl 1/4 teaspoon powdered saffron, mix with 5 tablespoons boiling juice from the saucepan. Mix thoroughly and add to boiling juice. Put around the fish 1/2-inch slices of French bread. Pour over fish. Serve piping hot.

Very simple *hors d'oeuvre* to precede the *Bouillabaisse*—neither vegetable nor fruit juice, please—but raw baby artichokes, endives washed and cut in half, radishes and asparagus tips for example; with coffee to end a perfect lunch.

We went to Marseilles to spend the day two or three times a month and had a *Bouillabaisse* at the best of restaurants.

We stayed on at Saint Rémy; summer was over, autumn was even more beautiful. If the *mistral* howled it not only made the sky bluer but all the landscape more vivid. One day we walked to a small Gothic chapel. In a bare field there was a single very large leafless and symmetrical Japanese persimmon tree heavily laden with its deep-orange fruit, silhouetted against the brilliant sky. It remains one of the loveliest of memories.

All day and all night we heard the sheep bells as the flocks were driven into the hills for the winter grazing. We were obliged to take the small roads. The flocks made an effective barrier on the main ones. Janet comfortably installed in her home could not understand why we didn't find Saint Rémy thoroughly insupportable. The food at the inn even for the Christmas and New Year's celebrations was wretched. The only thing the little town produced, and that was first rate, was *glacé* fruit, but one could not live upon that alone. French *glacé* fruit differs from our crystallised kind. In France the syrup in which the fruit is cooked is not boiled long enough to crystallise. In Saint Rémy there was a manufactory—if anything as unpretentious and small could be called that—of these fruits. They made a speciality of whole *glacé* melons filled with the smaller fruits, cherries, apricots, plums and pears, delicious and attractive. They sent them to us in Paris until the outbreak of this last war. Like so many other good things, they disappeared with the catastrophe.

As winter wore on, we became restless. Perhaps it was too much *mistral.* It was foolish to leave before the spring came. One day in March, walking through a ploughed field, we were forced to admit that the climate was no longer bearable and that it would be best to leave at once. We took several days to have a last long look at all the places we so little wished to leave and then drove back to Paris.

It had become our habit to remain in Paris during the winter and only take short trips in spring, which however did not prevent us from discussing projects for the long summer vacations. At this time a series of booklets on the gastronomic points of interest in the various regions of France were being published. As each one appeared I would read it with curiosity. The author was paradoxically a professional *gourmet.* Of the places we knew I was not always in agreement with his judgment. However, when it became time to plan the route we were to take to meet the Picassos at Antibes we chose one based on the recommendations of the guides. The first of them was Bourg-en-Bresse.

In May we started off for Chablis, where we would find not only incomparable food but my favourite wine, Chablis. Monsieur Bergeran was an intelligent and gifted *chef.* His menus were a history of the French kitchen and he was its encyclopaedia. One day he said to us that a true *chef* should have no secrets, that anyone could know everything there was to know about cooking. There should be no tricks or secrets. To prove this he asked me to come and see him at work. This is the way I saw him prepare

CHICKEN SAUTÉ AUX DUCS DE BOURGOGNE

Cut a fine roasting chicken into six pieces. Brown them in 4 tablespoons butter over medium flame in Dutch oven. When browned add salt and pepper, cover and put in preheated 350° oven. Baste frequently and turn once. It will take between 3/4 hour to 1 hour to cook according to size of chicken. When the chicken is cooked remove from flame and place the pieces of chicken on preheated carving dish. Put the Dutch oven over medium flame and add 3/4 cup port, 1/2 cup brandy, 1/4 cup whisky and 1/4 cup best kirsch. Detach from oven with spatula any glaze that may be adhering to it. Mix well. In a bowl stir 2 yolks of eggs and slowly add 2 cups warm cream. Pour slowly into Dutch oven. Heat thoroughly but do not allow to boil. Tip in all directions. Do not stir. Pour over chicken and serve hot.

From Chablis we went to Dijon, where we had for dinner at the famous restaurant of the Three Pheasants

SUPRÈME OF PIKE À LA DIJONAISE

Cut the fillets from a pike, see that no bones adhere and then skin them. Interlard them as one does fillet of beef. Put them in a deep dish with 1/4 cup brandy, 1/2 cup sherry and 1 cup good dry red wine, salt and pepper and 4 shallots chopped fine and 4 bouquets each containing 1 stalk of celery, 1 small twig of thyme and 1/4 laurel leaf, each bouquet tied in a muslin bag. Baste with liquid and put aside. In winter keep for 48 hours, in summer for 24 hours, basting twice a day. When the fillets are ready to be cooked place in a deep earthenware dish which has been heavily coated with soft butter, the fillets, the four little bags and the strained marinade. Put into preheated oven 400° for about 20 minutes, basting frequently. When the fillets are well browned, remove from oven, add 2 tablespoons cream and 3 tablespoons soft butter. Baste and serve at once.

From Dijon our road led to Bourg-en-Bresse, recommended by the Guide Gastronomique, through the country renowned for its chickens of the large and thick breasts and short legs. Bourg is a well-known market town, not only for its fowl but for dairy produce and vegetables. We fell under its spell at once. The menu at the hotel for dinner was carefully chosen and delicately cooked. We were delighted and toasted the guide book which had led us there and decided to stay for a couple of days. In the morning we visited the market and the provision shops, regretting that we did not live in the region—not suspecting that we were to spend six months a year for seventeen consecutive years within twenty-five miles of it. For lunch we had

TURNOVERS WITH CRAWFISH—SAUCE NANTUA

Make a puff paste, roll in a ball and put aside in a cool place. Dice 2 medium-sized carrots, 2 medium-sized onions, cut 1 stalk of celery in rings. Put 2 tablespoons butter in saucepan and when melted brown the vegetables in it. When browned, put into the saucepan 1 cup dry white wine and 48 unshelled crawfish. Cover and cook over medium heat for 10 minutes. Remove the crawfish with a perforated spoon. Remove their shells. Put the crawfish aside, and the shells and the diced vegetables into a mortar and pound until fine enough to strain through a fine sieve. Add 1 cup thick cream sauce, the strained juice in which the crawfish have cooked, 2 tablespoons soft butter, 1 teaspoon cognac. Mix well and cool. Roll out the puff paste 1/4 inch thick. Cut in squares of 4 inches and put 4 crawfish towards one corner with 1 tablespoon sauce which must be cold. Turn the opposite corner over to cover the crawfish. Press around the edges firmly with the floured handle of a small knife. Paint lightly with a stirred egg and put on lightly buttered baking sheet and bake in preheated oven 400° for 20 minutes. While the turnovers are baking keep the rest of the sauce hot. When the turnovers are baked remove from oven and place on preheated plate, and pour some of the sauce on each one of them.

These turnovers are very popular in the Bresse and the Bugey for Sunday lunch, for baptisms and for weddings.

From the same source the advice that had taken us to Bourg took us to Belley. The short ride was through a pleasing country. Belley was a small town on a hillside with varied landscape on all sides. For our evening meal we ordered simply a poached fish with brown butter, a vegetable salad and raspberries. It was satisfactory. We took a walk round the outside of the town and were enchanted. Though not actually in the mountains Belley had mountain air from the not far distant Alps. The next morning early we drove in Godiva in all directions. The country was beautiful and diversified. The people on the roads and in the fields were upstanding and had an air of well being. The children were charmingly pretty. In the hills there were lakes and in the valleys there were streams. It was too good to be true. We got back to the hotel for lunch. It was only then that we remembered how enthusiastically the cooking at the hotel had been recommended. The menu was commonplace, the cooking undistinguished. Even so we would stay on to see more of the country. We wired to the Picassos that we were delayed, we would get to them within a week. The proprietress of the hotel—her husband was the cook who preferred reading Lamartine in a corner to doing his work in the kitchen—when she heard we were staying on moved us to larger rooms with a view over a garden with the hills in the background.

Everything had been so much to our taste that we had been indifferent to the cooking. It was mediocre and would probably not improve. For the moment at least this was of no importance. At the end of the week when we drove over to Aix-les-Bains and had an unpretentious well-cooked lunch at a restaurant by the lakeside we could laugh at the cooking of the hotel at Belley.

Gertrude Stein wrote to Picasso that we were not going south this year, we found Belley sympathetic and were spending the summer there.

One evening a beautiful woman sat at the little table next to us, her book turned towards us. After several days she suddenly asked, half turning her head, if we had Lavaret for dinner every night, to which we answered that we did. Several evenings later in the same fashion she asked why we had Lavaret every evening for dinner. Gertrude Stein told her that it was the most carefully prepared dish on the menu. Obviously, said the lady. One day we met her in the street and we stopped to talk. She told us of several restaurants in small towns, even in villages, in the region where we could eat extremely well. Godiva took us to all of them. We had not after all lost our taste for food. We went to the *Haute Savoie* and on the lake of Annecy had remarkably good lunches, to Artemarre which pleased us even more, and to Saint Genix where in a simple *décor* we enjoyed uncommonly good cooking. But our favourite restaurant was totally unlike these, both for its quality and its simplicity. It had been a *bistro* before Madame Bourgeois for family reasons had taken it over. Within a short time it had become known by French *gourmets* as one of the best restaurants in France. Madame Bourgeois was a perfect cook. To the simplest dishes was given as much attention as to the most complicated, which occasionally included the great dishes created in the last three centuries. We drove over often to Priay, a village of 341 inhabitants near the river Ain which is known for its superior fish, in a hunting district abounding in pheasants, partridges and grouse, and within easy driving distance of the great markets of Lyon. Monsieur Bourgeois was a great judge of wine and went each autumn to the annual auction at the Hospice at Beaune, returning with its best. We got to know the Bourgeoises very well and always spent a moment with them in the kitchen, which was thoroughly organised and equipped and very nearly noiseless. From Madame Bourgeois I learned much of what great French cooking was and had been, but because she was a genius in her way, I did not learn from her any one single dish. The inspiration of genius is neither learned nor taught.

After the vintage it had turned cold and we went back to Paris. By April we had returned to Belley to the same pleasant hotel and its poor food. Friends came to stay with us and we would drive them to lunch at Aix-les-Bains, Artemarre and Annecy, but above all to Priay.

Before the end of the summer we realised that we must either build, buy or rent a house somewhere in the country near Belley. But that was a large order. We spent two summers at it. The land we wanted to buy was either not for sale or had no water, the houses were not for sale or had little water. There were no houses to rent that we would have moved into. We were miserable until one afternoon we glimpsed the perfect house from across the valley. It was neither for sale nor to rent but this time nothing would prevent our securing the summer home of our dreams. It was let to an officer in garrison at Belley. How did one dislodge a tenant without a legal reason? We talked to the owner of the house who plainly showed he considered us quite mad, but he told us that his tenant was a captain, and that there were too many majors in the battalion. That was enough to inspire us. We would get two influential friends in Paris to have him promoted, he would be ordered to another garrison and the house would be free for us. Soon after our return to Paris the captain came up to Versailles to take his examinations. He failed to pass them. Our friends said we were not to worry, in three months he had the right to try again. And once more he failed. We were despondent. Someone suggested his being appointed to Africa, at advanced pay and tantamount to promotion. The captain accepted, the friends became active again and soon we were ecstatically tenants of a house which we had never seen nearer than two miles away.

Godiva was tired and old and Gertrude Stein in spring bought a new car and we drove down to Bilignin in it with a white poodle pup to find the house better than our dreams of it.

VII.

TREASURES

WHAT IS THE FIRST FOOD YOU REMEMBER, REMEMBER SEEING IT if not eating it? Well, the first food I remember from my early childhood in San Francisco in the early 'eighties was breakfast food: cracked wheat with sugar and cream, corn meal with molasses and farina with honey. But after that the first food that I saw and clearly remember was *soufflé* fritters which of course were not included in a diet prescribed for a child. Nora, my mother's cook, fortunately stayed on long enough for me to taste her fritters. Nora left my mother's kitchen when she was nearly forty years old to marry a well-paid workman and she proceeded to produce five or six children. Maggie, the nurse, would go to see her and on her return would tell the incredible story that Nora who had been such an exquisite cook was now feeding her family, including the youngest born, on canned food. She was a precursor. This is the way to prepare

NORA'S SOUFFLÉ FRITTERS

Put 1 cup water, 1/2 cup butter and a pinch of salt into a saucepan on the fire. When the butter has melted and the mixture is about to boil remove from the fire and quickly stir in 2 cups sifted flour. Stir vigorously with a wooden spoon. Place on a very low flame until the mixture leaves the saucepan dry. Turn into a large bowl. Cool for 10 minutes. Then continuing to beat vigorously add 8 eggs, one at a time, thoroughly incorporating each egg before adding another. While beating lift the mixture high in the air so that as much air as possible will enter. Do not stir but beat steadily for about 20 minutes. Put aside in a cool place, but not in the refrigerator, for 2 or 3 hours. When ready to use heat sufficient oil for deep frying to medium heat. You will have rolled the mixture into balls the size of a large walnut in the centre of which you will have placed 1/2 teaspoon currant jelly. Increase the heat. The fritters will rise to the surface, swell, turn over and become golden brown without your aid. At this point remove them at once from the oil, coat generously with confectioner's sugar and serve immediately.

Nora's ice cream is still remembered. It was frozen in a then lately invented "automatic" freezer, there was no cranking to be done. My mother revelled in each year's invention. How she would have enjoyed the present gadget of the week. I found the directions for the use of the "automatic" freezer pasted in the back of the cook-book she had bought for Nora. Many years after, Maggie told my mother that Nora was completely illiterate and so she had read aloud to Nora the recipes she needed and had written down the daily expenses every evening. Maggie had not wanted my mother to know this while Nora was in her service, she thought it would embarrass both of them.

NORA'S ICE CREAM

1 quart whipped cream sweetened with 3/4 cup icing sugar. Add 1 1/4 cups raspberry jelly slightly melted. Fold in the beaten whites of 5 eggs. Freeze.

It surprises me, recalling these two of Nora's delights, to find that my collecting of treasures commenced so very, very long ago and that many of them, consequently, are no longer treasures. One must have been more innocent and more inexperienced than one likes to think of one's self as having been. If taste is a matter of choice, the quantity of rejections for this book is neither flattering nor encouraging. The wastepaper basket is too small. But if there are amongst the discards proofs of an undiscerning past, there will also, I hope, be signs of more recent perspicacity in those that are offered here.

My collecting of treasures began with Sweets—cakes and ices, desserts and candies—double evidence of their date. For when one is young, that is what interests one most; the rest follows. So this is one of the earliest.

SCHEHEREZADE'S MELON

Cut a piece from the stem end of the melon. Scoop out in as large pieces as convenient as much of the pulp as is possible without piercing the melon. Empty all the juice, dice the pulp in equal quantities (this will depend upon the size of the melon) as well as pineapple and peaches. Add bananas in thin slices, and whole strawberries and raspberries. Sugar to taste. When the sugar mixed with the fruit has dissolved put the fruit and their juice in the melon. Cover with four parts very dry champagne and one part each of Kirsch, Maraschino, Crème de Menthe and Roselio. Put in refrigerator overnight.*

This dessert and a complicated Bavarian cream which had a similar flavour from its including the same fruits and precisely the same liqueurs and cordials were early favourites. The recipe for Scheherezade's Melon is preserved in my mother's handwritten cook-book.

Another early recipe has for the last sixty years been known amongst my friends as

ALICE'S COOKIES

On a floured board sift 2 cups flour. You will require 2 1/2 cups unsalted butter, the yolks of 6 eggs and 1 cup icing sugar with which a vanilla pod has been pounded and sifted. With the tips of the fingers very lightly work in one-sixth of each of the ingredients until all of each of them has been worked in. It may take

* *Note.* Roselio is a Catalan liqueur or cordial which is also made round about Perpignan. It has not been imported into England since the war: Grenadine might replace it.

more flour, depending upon the size of the yolks of the eggs and the quality of the butter. Only add enough flour to roll. Roll to about 1/4-inch thickness. Cut with a round cookie cutter of any size that suits, but not more than 2 1/2 inches in diameter. Place on cookie sheets and bake in 300° oven for about 1/4 hour. The cookies should not be coloured. When done remove very carefully from cookie sheet with a metal spatula. They are as fragile as they are exquisite. Cover generously with sifted icing sugar. Do not put in tin box until cold. If the box has a cover that closes hermetically, the cookies will keep for two weeks or more.

As these cookies take some time to prepare, eventually they were replaced by an adaptation of Scotch short bread.

NAMELESS COOKIES

Cream 1 cup butter, very slowly add 1/4 cup sifted icing sugar and about 2 cups sifted flour. It should take about 20 minutes' active stirring. Do not beat but stir. When about half the flour has been worked in add a tablespoon best white curaçao and 1 teaspoon brandy. The dough should only be stiff enough just to hold its shape when rolled by one's extended fingers into small sausages a thumb's length and width. Place on cookie sheet an inch apart—they will spread in the oven, which should be preheated at 275°. The cookies should be very pale, indeed not coloured at all by the heat. Twenty minutes' baking will be sufficient. Remove carefully with metal spatula from cookie sheet and sieve icing sugar over them at once. In a well-covered tin box they will keep three weeks. These cookies have a less delicate texture than the Alice's cookies but are a time-saving substitute.

This recipe is one that was prepared at my request for a lunch celebrating a birthday to which some of my schoolmates had been invited. It is named after a cook.

KATIE'S CAPON

Brown a capon in 6 tablespoons butter in an enamelled pot over medium heat. Cover and reduce heat. After 1/4 hour add 3/4 cup hot water, 3/4 cup hot port wine, the zest of 1/2 orange, 1 teaspoon salt, 1/4 teaspoon pepper and a pinch of cayenne pepper. After 20 minutes, baste every 1/4 hour. The capon will be cooked in 3/4 hour. Remove from pot, place on heated serving dish. Skim juice in the pot and strain. Replace over lowest flame. Stir into pot 3/4 cup heavy cream. Add 4 tablespoons butter. Do not allow to boil, do not stir but tip pot in all directions. Serve very hot. The gravy is poured over the capon so that the juices of the capon when carved enter into the sauce. It is equally delicious cold. To serve cold the juice must have been thickened with rice flour or arrowroot—as in those days one did for more delicate food—and cooled before pouring over the capon.

When treasures are recipes they are less clearly, less distinctly remembered than when they are tangible objects. They evoke however quite as vivid a feeling—that is, to some of us who, considering cooking an art, feel that a way of cooking can produce something that approaches an aesthetic emotion. What more can one say? If one had the choice of again hearing Pachmann play the two Chopin sonatas or dining once more at the Café Anglais, which would one choose?

To return to our muttons.

LEG OF MUTTON À LA MUSCOVITE

Chop 3 large onions, 3 large carrots and 3 heads of celery. Brown them lightly in 4 tablespoons butter in a saucepan over medium heat. Add 1 cup boiling water and 1/4 cup boiling vinegar, 1 bay leaf, 2 cloves of garlic, 3 cloves, 1 teaspoon salt and 1/2 teaspoon pepper. Cover, reduce heat and boil for 20 minutes. Remove from heat and when cold, strain. Place the leg of mutton in a preheated

475° oven and after 20 minutes commence to baste with hot juice of vegetables. It will take about 1/2 hour to roast, according to the size of the meat.

MUTTON CHOPS IN DRESSING-GOWNS

Cook 6 mutton chops in a covered saucepan over medium heat in a hot *bouillon* that scarcely covers, with very little salt and a bouquet of the usual herbs. When they are cooked, or in about 15 minutes, remove chops from *bouillon*. Return *bouillon* to medium heat and reduce to a glaze. Then replace chops in the saucepan and glaze them. Remove from saucepan and spread on each chop this mixture: finely chopped parsley, 1 finely chopped shallot, 2 chopped hard-boiled eggs, 1/4 lb. finely chopped mushrooms which have been cooked for 8 minutes in 2 tablespoons butter and 1 teaspoon lemon juice, 1/4 teaspoon salt and a good pinch of pepper. These ingredients are bound together with 3 tablespoons heavy cream. Paint the mixture on the chops with melted butter, dip in fine fresh breadcrumbs. Place for 15 minutes in a preheated 500° oven. Serve very hot.

In my collection of recipes there is one, *Rosbif de Mouton*, from a manuscript cook-book lent to me by a French friend. *Rosbif* is what the French call roast beef. It reminds one of the signs one used to see in Paris at smart tea shops—*Le Fif o'Clock à Toutes les Heures* (the five o'clock [tea] at all hours). The recipe for the Roast Beef of Mutton is by no less a person than Alexandre Dumas, senior, author not only of the *Three Musketeers* but of *The Large Dictionary of the Kitchen*. This recipe is entirely devoted to the manner he recommends for skewering the hind half of a sheep that is to be roasted on the spit. For this reason it is not given, but there are in my collection two other of Dumas' recipes. They too are for the preparation of mutton. One is for

FOIE DE MOUTON À LA PATRAQUE
(Gimcrack Mutton Liver)

Put a sheep's liver cut in thin slices into very hot olive oil. Cook each side of the slices for 5 minutes. Remove the meat and add to the oil the juice of 2 lemons or the same amount of vinegar, garlic and very finely rolled toast. Mix these well together for 2 minutes and return the meat and some chopped parsley, and *sauté* until the liver is cooked. Serve hot.

It is dated by the use of toasted crumbs to thicken sauces; in fact, though, they were little used as late as Dumas senior's lifetime. This is the other Dumas recipe:

THE SEVEN-HOUR LEG OF MUTTON

In an earthenware pot place the rind of pork fat cut in small pieces. Interlard a leg of mutton with ham, garlic and lard. Put your leg of mutton into the pot with salt, pepper, 2 large onions, 3 glasses water, 1 glass white wine. Cover the pot with a plate and paste paper around the pot and the plate. In the plate pour some wine and allow it to simmer for 7 hours.

From the same rich fund that produced the two Dumas recipes there are others to offer, beginning with

CHICKEN À LA REINE MARIE

In a casserole over medium heat place 5 tablespoons butter and 1 1/4 lbs. veal knuckle cut in four pieces. Let them brown lightly on all sides and remove from casserole. Do not allow the butter to burn. Add 4 large onions cut in thick slices and brown them. Remove from casserole. Put 4 large carrots cut in thick slices in the casserole and brown them. Add the pieces of veal knuckle and the onions. Reduce heat and let them cook together for 1/4 hour, add 1/2 teaspoon salt and 1/4 teaspoon pepper. Add a calf's foot cut in four pieces. Cover with 6 cups hot water and add 1 large bouquet. Cover the casserole and over lowest flame simmer for 4 hours. Watch carefully that the water does not completely evaporate. Then remove the veal knuckle, the calf's foot and the bouquet. Mash well through a sieve so as to extract all the juice the vegetables contain. Melt 4 tablespoons butter in a saucepan over low flame. Add 1 tablespoon flour and, constantly stirring with a wooden spoon, let it brown lightly. Very slowly add the strained juice. Let this sauce simmer until thickened. Cut a fine chicken into joints and place in an iron pot in which 7 tablespoons butter have been melted. Over medium heat brown all sides of the pieces of chicken. Add 1/2 teaspoon salt and 1/4 cup good dry white wine. Cover and simmer for about 3/4 hour, depending upon the size of the chicken. The giblets will have been cooked slowly for about 2 1/2 hours with 1 carrot, 1 turnip and 1 leek. This should produce 1 1/2 cups concentrated *bouillon.* In a small saucepan melt 4 tablespoons butter over low heat. Add 1 tablespoon flour. Stir with a wooden spoon and mix thoroughly. Add the *bouillon* and when it commences to boil slowly pour in 1 3/4 cups heavy cream. Add 1/2 teaspoon salt and 1/4 teaspoon pepper. Bring to the boil and remove from heat. Arrange the pieces of chicken in a manner to resemble an uncut chicken on a serving dish, and when the sauce is tepid pour over the chicken. Serve cold.

This is a very satisfactory principal dish for hot weather. A green salad accompanies it.

From the same source,

A SALMON PÂTÉ

For the crust, sieve 4 cups flour in a bowl with 1 3/4 cups butter, 1/4 teaspoon salt and mix with a pastry blender. Add only enough cold water to hold the dough together. Do not mix it too much. Put aside covered for 1/2 hour. Then roll it to a square of about 12 inches. Fold it from top to centre and from bottom to top. Fold the sides in the same manner. Put aside covered for 10 minutes. Roll and fold a second time. Put aside covered. (The dough can be rolled in wax paper in these modern days instead of covered, as this old recipe recommends.) Boil 2 eggs for 11 minutes. Boil 1 cup rice in 3 cups water for 15 minutes. Dry thoroughly in the oven. Roll the dough to a 1/4-inch thickness, then roll all around the edge to half that thickness. This will leave the centre with a double thickness. In the centre spread a layer of rice (it must be perfectly dry and cold) of about a 1/3 inch in thickness, salt and pepper. Cover with a layer of heavy cream, 1/3 inch in thickness. Then place on the cream 1 lb. raw salmon cut in three slices, let them overlap, salt and pepper them. Cover with another layer of heavy cream 1/3 inch in thickness. Sprinkle the 2 hard-boiled eggs finely chopped over the cream. Cover with a layer of rice of 1/3 inch in thickness, salt and pepper lightly. Paint the uncovered dough lightly with water and cover the filling from the top and bottom. Then from the sides. Paint the top and the sides with an egg mixed with 3 tablespoons water. Bind the sides of the *pâté* with a piece of buttered paper and tie with a string to hold it in place. Prick the top of the *pâté* with a fork in a number of places. In the centre of the top cut a small round opening of 1 inch in diameter, remove the small piece of dough. Slip the *pâté* on to a lightly buttered baking sheet and place in a 500° preheated oven. Lower the heat in 10 minutes to 400°. As soon as the *pâté* commences to colour, lower to 375°. In 1/2 hour remove the paper that binds the sides and bake for 1/4 hour longer. Remove from oven, pour into the hole in the centre 4 tablespoons melted butter and serve.

The recipe for this omelette comes from the cook-book of Georges Sand. Should it not therefore be called

OMELETTE AURORE

Beat 8 eggs with a pinch of salt, 1 tablespoon sugar and 3 tablespoons heavy cream. Prepare the omelette in the usual manner. Before folding it, place on it 1 cup diced candied fruit and small pieces of *marrons glacés* which have soaked for several hours in 2 tablespoons curaçao. Fold the omelette to keep the fruit in place, on fireproof serving dish. Surround with *marrons glacés* and candied cherries. Cover at once with this Frangipani cream, made by stirring 2 whole eggs and 3 yolks with 3 tablespoons sugar until they are pale-lemon coloured. Then add 1 cup flour and a pinch of salt, stirring until it is perfectly smooth. Add 2 cups milk and mix well. Put in a saucepan over lowest heat and stir until quite thick. It must not boil. Be careful that the cream does not become attached to the bottom or sides of the saucepan. When it has thickened remove from heat and add 2 tablespoons butter and 3 powdered macaroons. Stir and mix well. Pour over omelette and sprinkle 1/4 cup diced angelica over the cream. Then sprinkle 6 powdered macaroons on top and 3 tablespoons melted butter. Place the omelette in a preheated 550° oven only long enough to brown lightly.

OMELETTE IN AN OVERCOAT

Put in a saucepan over medium heat 3 tablespoons butter with 1 tablespoon chopped parsley, 3/4 cup mushrooms cut in thin slices, stems included, 2 chopped spring onions and 2 shallots finely chopped. Stir them until they are coated with butter, then add 1 1/2 tablespoons flour and 1 1/2 cups milk. Continue to stir until the sauce thickens. Lower heat after it comes to a boil and simmer for 5 minutes. Pour into the serving dish. Place on it 6 mellow eggs, which are eggs placed for 4 minutes in boiling water, removed and placed under the cold-water tap, and when they are cold carefully shelled. They are then placed in a saucepan of boiling water, covered and allowed to stand in it for 5 minutes. Prepare a 6-egg omelette in the usual manner, entirely cover the eggs and sauce with it. Sprinkle the top with 2 tablespoons melted butter, then a thick covering of finely toasted breadcrumbs and once more a sprinkling of 2 tablespoons melted butter. Brown in preheated 550° oven. Serve at once.

And one more omelette, but without a name, so it will be called

OMELETTE SANS NOM

Stir in a bowl the yolks of 4 eggs with 3/4 cup heavy cream, a pinch of salt and 1 tablespoon grated Parmesan cheese. Mix well and add 5 tablespoons very soft butter. Beat the whites of 4 eggs and fold the egg-cream mixture into them at the same time as 1 1/2 cups tiny dices of ham. Put in a preheated fireproof well-buttered dish for serving and place in a preheated 500° oven for 5 minutes. Remove from oven and quickly cover with 1 cup heavy cream on which 1/4 cup tiny diced ham is sprinkled. Return to oven, lower heat to 400° and brown for a little less than 1/4 hour more. Serve at once.

ROAST KIDNEYS

Cook in a frying pan over medium heat a veal kidney. Cut into small pieces 6 tablespoons butter. Stir until all the pieces are covered with butter. Add salt, pepper, 1/2 cup hot Madeira. Lower heat and cover. Cook for 6 minutes. Remove pieces of kidney with a perforated spoon and put aside. Add to the frying pan 2 tablespoons butter and brown lightly on each side six slices of bread and remove from pan. Put kidney through the meat chopper. Add 7 tablespoons butter, 1 tablespoon chopped parsley, 1 chopped shallot, 1/2 teaspoon salt, 1/4 teaspoon pepper and the yolks of 4 eggs. Mix well. Beat the 4 whites of eggs and fold into the chopped mixture. Place the browned bread on a fireproof serving dish, put a layer of the mixture on each piece and flatten the top with a knife. Cover with fine breadcrumbs and melted butter. Place in preheated 400° oven for 20 minutes. Serve with this

RAVIGOTE SAUCE

Put in a small saucepan over low heat 1 cup of strong *bouillon*, 1 teaspoon vinegar, 1/3 teaspoon pepper, 1/4 teaspoon salt, 1 tablespoon butter mixed with 1 teaspoon flour. Stir until it commences to boil. Allow to simmer for 5 minutes. Then mix 1 teaspoon chopped chervil, 1 teaspoon tarragon, 1 teaspoon chopped capers, 1 teaspoon chopped parsley, 1 teaspoon chopped shallot and 1 crushed clove of garlic. Put in a mortar and pound well. Then add them to the sauce and heat. Stir until the sauce is green. Do not boil. If the herbs are not finely enough powdered—which would be a mistake and a pity—strain and reheat in another saucepan but do not allow to boil. Serve at the same time as the kidneys.

In the collection of recipes there are certainly a disproportionate number for the preparation of chickens, but the French when in doubt always answer chicken. This is a very witty way to present a

QUADRIPARTITE CHICKEN

Cut a fine chicken in four pieces. Cover each piece completely with a thin slice of fat back of pork and skewer or tie to keep in place. Put them in an iron pot over medium heat with 4 tablespoons butter, brown very lightly and add 1 truffle, 1 large slice of ham, a bouquet of parsley, a twig of thyme, 1/2 laurel leaf and several leaves of basil or 1/4 teaspoon powdered basil, 1 clove, salt, pepper and 3/4 cup dry white wine. Simmer covered for 3/4 hour. Chop the truffle, the slice of ham, the yolk of 1 hard-boiled egg and 10 capers, each of them separately chopped. Remove the pieces of chicken to the serving plate placing each piece to reconstruct the chicken, but flat. Skim the sauce through a strainer and replace over heat. Bring to a boil, stir the bottom and sides of the pot and add 1 1/2 tablespoons butter and 1 tablespoon flour which have been thoroughly mixed together. Stir and boil gently for 3 or 4 minutes until the sauce is thickened. Pour over the four pieces of chicken. Cover one with chopped ham, another with yolk of egg mashed through a sieve, another with chopped truffle and the last one with chopped capers.

This recipe for rice taught me how to judge when rice was sufficiently cooked. Besides this it is an excellent dish.

RICE À LA DREUX

Wash, drain and thoroughly dry 1 1/2 cups rice. Put on a saucepan over medium heat and melt 3 tablespoons butter. Add the rice and stir with a wooden fork until all the rice is covered with the butter. Lower heat and continue to stir for 10 minutes. Then add 3 cups chicken broth, 1/2 teaspoon salt and 1/4 teaspoon pepper and cover. The rice is cooked when small holes appear on the surface. Put the saucepan uncovered in a very slow oven to dry the rice. Cut a veal kidney in thin slices, put in a saucepan in which 3 tablespoons butter have been melted. Over medium heat, brown lightly and add 1 tablespoon flour. Mix well and add 3/4 cup hot chicken *bouillon* and 1/4 cup hot Madeira. Reduce heat as soon as it commences to boil. Add 1/4 teaspoon salt, 1/8 teaspoon pepper and a pinch of powdered mace or nutmeg. Cover and allow to simmer for 10 minutes. Scramble 8 eggs in saucepan in which 2 tablespoons butter have been melted, add 1/4 teaspoon salt and 1/2 cup cream. Be careful not to overcook. The rice having been placed in a ring mould and tapped so that there will be no holes is now removed from mould on to the serving dish. The centre of the dish is filled with the scrambled eggs and the slices of kidney are placed around the rice and their sauce poured over them.

MUSHROOM SANDWICHES (1)

Mushroom sandwiches have been my speciality for years. They were made with mushrooms cooked in butter with a little juice of lemon. After 8 minutes' cooking, they were removed from heat, chopped and then pounded into a paste in the mortar. Salt, pepper, a pinch of cayenne, and an equal volume of butter were thoroughly amalgamated with them. Well and good. But here is a considerable improvement over them, also called

MUSHROOM SANDWICHES (2)

The method is the same as above up to a certain point. These are the proportions. For 1/4 lb. mushrooms cooked in 2 tablespoons butter add 2 scrambled eggs and 3 tablespoons grated Parmesan cheese and mix well. The recipe ends with: This makes a delicious sandwich which tastes like chicken. A Frenchman can say no more. Which gave me the idea of introducing chicken sandwiches in which chopped and pounded chicken was substituted for the mushrooms. Naturally they were well received.

So we are back to chicken with some recipes for them, delicious or original. This one is both.

A HEN WITH GOLDEN EGGS

Put a hen in a saucepan over very high heat. It should be covered with cold water, and when it is about to boil, the water is skimmed. Then reduce the flame, add salt, pepper, a bouquet of laurel leaf and a sprig of thyme and parsley, 1 large onion, 1 large carrot,

1 leek and 2 large stalks of tarragon. It will depend upon the age of the hen whether it will be tender in 1 1/2 or 3 hours. It should not cook beyond being tender. Before the end, add 1/2 lb. mushrooms. Let them boil in the *court-bouillon* for 10 minutes. Remove chicken. If there is more than 3 cups *bouillon*, reduce and strain, add 1 cup heavy cream and 1 cup butter. Do not allow to boil. Do not stir but tip saucepan in all directions. In the cavity you will place the golden eggs that are made in this way. Boil 2 lbs. potatoes in their jackets. When tender, peel and put through the mechanical vegetable masher. Then beat hard while slowly adding 1/2 cup butter and the yolks of 6 eggs. Mould into balls and shape to imitate eggs, dip into a bowl of melted butter and cover generously. Put into frying pan in which 5 tablespoons butter has been melted over medium heat, heat thoroughly but do not brown. Stuff the cavity of the hen with these eggs. Pour the sauce over the chicken and surround with the rest of the golden eggs.

This is an amusing way to present a chicken. It is a delicious dish.

CHICKEN À LA COMTADINE

Cut a spring chicken into eight pieces, salt and pepper them. Lightly brown them in 4 tablespoons butter in a pan over medium heat. Lower heat and cover the pan, let them cook for 20 minutes, shaking the pan frequently. Remove from heat and pour over them the contents of the pan. Then pour into the pan 1/2 cup good sweet Italian vermouth (Cinzano or Rossi). Heat it and light it. While still lighted replace the pieces of chicken, extinguish flame. Stir to heat and add 1 tablespoon tomato jam, a good pinch of cinnamon, and one of cayenne pepper. Stir the chicken for 5 minutes to coat each piece with the sauce and serve.

And this fowl with wine—

DUCK IN PORT WINE

Put 24 figs in a wide jar to marinate for 36 hours in an excellent dry port wine and cover hermetically. Put the duck in a preheated 450° oven. After 1/4 hour commence to baste it with the port wine in which the figs have been macerating, and which has been heated. Continue to baste every 15 minutes. Turn the duck on each side so that the legs are browned. When all the port has been used for basting put the figs around the duck and baste with veal *bouillon*. Continue to baste. The duck will be cooked in an hour unless it is a very old duck indeed.

No duck interests me after the month of September. Ducks should be goslings and it is a pity that they are not hatched later in the season, for they are really a late autumn and winter dish.

This is a very pretty dish and, though original, not outrageous. On the contrary, it is a satisfactory combination of colour and flavour.

PINK POMPADOUR BASS

Bass should not be cooked in a *court-bouillon*. Put a 4 lb. bass into hot salted water. As soon as the water boils again reduce at once to low flame. The fish should simmer. The water should tremble or shudder, as the French say, for 3/4 hour. Remove from kettle and delicately remove the skin, and when quite cold place on serving dish and cover with a pale-green mayonnaise made with 2 yolks of eggs, 2 1/2 cups oil, the juice of 1 lemon, salt and pepper and enough cress, chervil and tarragon leaves in equal quantities pounded in a mortar until reduced to a smooth paste to make in all 4 tablespoons. Boil 2 lbs. potatoes in their jackets, peel and put in the electric blender with half the volume of boiled beetroots. When they are mixed add 1 tablespoon oil and juice of 1/2 lemon, salt and pepper and enough thick cream to put in the pastry tube, and decorate around the fish.

And this is another attractive and tasty dish.

GIANT SQUAB IN PYJAMAS

Completely cover a giant squab (young pigeon) with a thin slice of back fat of pork. Tie it to keep in place, cover with a thick coating of butter. If possible to secure, cover the pork fat with vine leaves. Place in a *cocotte* with an onion, a crushed clove of garlic, a bouquet, 1/3 cup diced skin of back fat of pork, 3 tablespoons butter and pepper, and 1/3 cup brandy. Cover and simmer over low heat for 2 hours. Remove lard and place squab on serving dish. Prepare an aspic by soaking 1 tablespoon gelatine in 1/2 cup water. Bring 3/4 cup port wine and 3/4 cup chicken *bouillon* to a boil. Melt the gelatine in the hot mixture. When it is cold and sufficiently thick, cover squab. Shred 1/2 red cabbage, season with 1/2 teaspoon salt, 1/4 teaspoon pepper. Add 1 1/2 tablespoons olive oil and the juice of 1 lemon. Mix well and surround the squab with this salad. Surround the squab with a few nasturtium flowers and place a few around the edge of the salad.

PHEASANT WITH COTTAGE CHEESE

Fill the cavity of a pheasant with cottage cheese. Sew the cavity together securely. Tie round the pheasant a thin slice of fat back of pork and paint generously with melted butter. Put 3 tablespoons butter in a *cocotte* over medium heat. When the butter is melted place the pheasant in the *cocotte* and reduce heat. Add salt and pepper. Cover and cook over low heat for 1 hour, basting with the melted fat. One-quarter hour before the pheasant is cooked, remove the pork fat and brown the pheasant in the fat on all sides.

After placing the bird on its serving dish, add 1/3 cup brandy to the *cocotte*, scrape the bottom and sides of the *cocotte* to mix the glaze that will have adhered. Pour over pheasant and serve.

MUSSELS WITH A CREAM SAUCE (1)

Put in a fish kettle 2 quarts thoroughly scrubbed and washed mussels, 3 tablespoons butter, 3 chopped shallots, 1 clove of crushed garlic. Cover and over highest heat cook for 4 or 5 minutes. As soon as the shells open, the mussels are cooked. Drain the mussels. Strain the juice in which they are cooked through a fine hair sieve. Remove the mussels from the shells. Pour the juice into a bowl. There will be a deposit in the bowl. Be careful that it is not poured into the saucepan. Put saucepan over low heat, and when the juice is hot pour a small quantity of it into a bowl. Add 1/2 tablespoon saffron and mix well. Add this to the juice. Add 2 tablespoons flour, stir until the sauce has thickened, about 5 minutes. Add mussels. Reduce heat and slowly stir in 3/4 cup heavy cream. When it is quite hot remove to deep serving dish, sprinkle with chopped parsley and serve.

The mussels can be served cold with a

Fennel Sauce

made by adding 1 whole fennel cooked covered in boiling salted water for 3 minutes, removed from water, drained, pressed and wiped dry. Then chop very very fine and add to a Sauce Mousseline.

If one likes mussels at all one likes them madly, and that is the reason there are so many recipes for their preparation amongst the treasure. Here is another and last one:

MUSSELS WITH A CREAM SAUCE (2)

Put in a saucepan over medium heat 3 tablespoons butter. When it is melted, add 1 large carrot cut in thin slices. Reduce heat, cover and cook for 1/2 hour, shaking the saucepan from time to time. Then raise heat to high, add 3 quarts scrubbed and washed mussels, 1/4 teaspoon salt and 1/4 teaspoon pepper. Cover and cook for 4 minutes. Remove mussels, drain and remove from shells. Put in a saucepan over medium heat 2 tablespoons butter and 1 1/4 cups chopped mushrooms. When they are hot, add 3 tablespoons flour. Stir with a wooden spoon. Add slowly the juice of the mussels which has been strained through a hair sieve. Stir with a wooden spoon. Boil for 4 minutes. Reduce heat and add 5 tablespoons butter and 1/2 cup heavy cream. Do not allow to boil and do not stir. Mix by tipping saucepan in all directions. Remove from flame; a few drops of lemon juice may be added, for those who care for it.

Most of the treasures in this collection are French and this is intentional. Though collecting began in the United States, my cooking days have been spent in France. There are however some Italian and Spanish treasures, which doubtless in the United States now belong to the public domain. But one Spanish recipe is tempting. It must be confessed it came to me from a French friend and is therefore unquestionably a modified version of the original. Chauvinism apart, how can one feel otherwise than that it has been contaminated through a natural prejudice. Here it is:

FISH IN A SPANISH PIE

Mix 1/2 cup olive oil, 1 egg, a pinch of salt, a good pinch of powdered anis, 1 1/2 tablespoons brandy. When thoroughly mixed, add 2 1/4 cups flour, work with a pastry blender but do not overwork. Add only enough water to hold the dough together. Put on a floured board, knead it out thin with the palm of the hand to about 1/2-inch thickness. Roll into a ball and repeat the rolling and kneading. Roll into a ball and put aside for several hours in a cool place (6 hours is recommended). The kneading must be done quickly. This amount of dough will be sufficient for a fish weighing 2 to 2 1/2 lbs. Spaniards prefer trout to other fish—they consume an incredible amount of it. So a trout it should be, but any fish that hasn't too strong a flavour will do. After the fish is cleaned, cut open down the stomach side. If one is not butterfingered, but oilfingered, with a sharp knife the backbone can be removed. In any case the fish is rubbed inside with salt and powdered nutmeg, painted with olive oil and stuffed with hazel nuts and seedless raisins in equal quantity. The nuts are placed in the oven to brown lightly. This helps to remove the skins by rubbing them in a dishcloth while still hot. Be sure that no skin remains. It is very unpleasant to encounter. Tie the fish together. Roll out the dough, place the fish on it, moisten the edges of the dough. Cover the fish completely. Press the edges of the dough together and place fish in a fireproof serving dish in a 450° oven for 35 to 40 minutes. It is a delicious dish.

The French make their version of pumpkin pie. It is a country dish, Paris does not know it. In the Bugey for the vintage they make one as we do in the United States, except for the omission of nutmeg or cinnamon flavour and of currants. When we told this to the daughter of a farmer who brought us one she had baked, she said they didn't grow spices, it was not hot enough, and their grapes were not suitable for drying and becoming currants. When it was suggested that nutmeg and currants could be bought in Belley, she was aghast and said that they never, but never, bought any provisions except coffee and sugar.

This is a pumpkin pie, but not a sweet one. It is called

THE CITROUILLAT

Take 5 lbs. pumpkin, peel it and cut it into 2-inch cubes. Place them in a bowl in layers with a sprinkling of salt between each layer. Let stand overnight. The recipe given above for the crust of Fish in a Spanish Pie is suitable if butter replaces the oil and the anis is omitted. In the morning wipe the cubes of pumpkin, and when dry place in a flat ovenproof dish, cover with the dough. Leave a 1-inch hole in the centre. The French like to decorate the opening by placing around it leaves cut from the dough. Place in preheated 400° oven for 3/4 hour. When it is removed from oven put into the hole in the centre as much heavy cream as the pie will accept, about a cupful, if the pie is inclined in all directions.

I did not think that corn meal was much used by the French but it is in the Jura, even in Burgundy. This is one of their recipes for a cake.

CORN-MEAL CAKE

(the French call it Croquants, which would be Crispies to Americans)

Mix thoroughly 1 cup corn meal, 2 cups flour and 1 cup sugar. Over very low flame slowly melt 1 cup butter and pour it over the dry ingredients. Mix thoroughly with the back of a spoon. Pat into a buttered pie plate with a detachable bottom. Place in a preheated 400° oven for 15 or 20 minutes. Be careful that it does not burn. It can be covered with a water icing to which 1 teaspoon rum has been added. Cut while hot.

And now for some recipes for desserts.

An English friend told me that her grand-aunt who was a lady-in-waiting to Queen Victoria bought a camel-hair India shawl and wore it to go to the palace. To her everlasting shame, a label tag which had not been removed and which was plainly visible was read by one of the royal dukes: Chaste but Elegant. It is a suitable name for this dessert,

RICE WITH FRUIT

Put 2 cups well-washed rice in a saucepan of cold water over high flame. Bring to a boil over medium heat 4 cups milk, 1 cup sugar and the rice. As soon as it comes to a boil, remove and drain. Put under the cold-water tap and drain. Reduce heat to lowest flame. Put rice in 1 quart milk and simmer for 1 hour without stirring. Place an asbestos mat under the saucepan. When cooked, add 1 teaspoon rum and pour into lightly oiled ring mould (preferably oiled with sweet almond oil). When it is cold, turn out and chill. The centre is filled with sweetened whipped cream. Decorate with strawberries. Cover the rice and keep enough to fill a bowl with this sauce: equal parts of strawberries and raspberries mashed through a fine sieve sweetened with icing sugar and flavoured with juice of 1/2 lemon and 1 teaspoon rum.

How or when this next recipe came into the cook-book of a French family they were not able to say, but it was written in a handwriting they recognised as that of a great-grandmother. It is for

CRÈME MARQUISE

(Jean-Jacques Rousseau was underlined)

For three servings, melt 3 ozs. chocolate over hot water. When it is melted add in small quantities at a time 8 tablespoons butter. Stir continuously. Add the yolks of 3 eggs, one at a time, stirring in the same direction. Remove from heat and fold into the stiffly beaten whites of 3 eggs. Chill before serving. It is really an excellent version of a common French dessert.

This too is a recipe of a common dessert, but a very good one, for

EMPRESS RICE

Wash 1/2 cup and 1 tablespoon rice in several waters, then soak in cold water for 2 hours. Drain and put in a saucepan of boiling water. When the water reboils, remove rice after 2 minutes and put it into 2 cups boiling milk with 3/4 cup sugar and allow it to simmer over low flame for 3/4 hour. Boil 2 cups thin cream with 1/4 cup sugar. Stir yolks of 4 eggs, mix with hot cream. Stir with wooden spoon over lowest flame until spoon is coated with mixture. Remove from heat, add 1 tablespoon vanilla extract and stir occasionally until perfectly cold. Then add 1 cup whipped cream. Gently mix into the rice 1 cup thinly sliced mixed fruits—bananas, berries, apricots and peaches that have been soaking in rum and in their juice, and 1/4 cup angelica cut in very small dice. Put into a lightly oiled mould in the refrigerator for at least 2 hours.

The French serve this dessert with a fruit or vanilla sauce, but perhaps we would prefer whipped cream.

An excellent ice cream is made with honey instead of sugar.

NOUGAT ICE CREAM

Heat 3 cups thin cream in saucepan over low heat. Stir 6 yolks of eggs and add the hot cream. Put over lowest heat and stir until spoon is coated. Remove from flame and stirring continuously pour it slowly over 1 cup honey (preferably orange flower). Add 1 tablespoon orange-flower water. Strain, and when cold incorporate 1 1/2 cups whipped cream and fold in 3 whites of eggs beaten stiff, 3/4 cup pistachio nuts that have been blanched, skinned and thoroughly dried, and 1/2 cup blanched almonds cut in half lengthwise (with the point of a knife they open very easily while still moist). Flavour with 1 tablespoon orange-flower water and freeze.

Here are four recipes, interesting at least for the names attached to them. To commence with one of Stephane Mallarmé's in his own words. He calls it marmalade, but it is an incomparable dessert.

COCONUT MARMALADE

No one who has entered and taken up from a counter a coconut and bought it knows what to do with it. For Parisians this classic fruit from afar, amongst the pomegranates or oranges and pineapples, remains a useless curiosity. But here is one of the most delicate dainties, of which it is the principal ingredient, from the islands and their coasts. Put 2 cups sugar and 1/2 cup water in a copper kettle and boil until it comes to the little pearl, drop in the grated coconut and stir with a wooden spatula. After 15 minutes, put 2 eggs into another kettle, pour the coconut into it, stirring in the same direction. Flavour with vanilla, cinnamon or orange-flower water, return to the fire for 5 minutes, and after letting it cool for 5 minutes pour it into a *compotier* (fruit dish) and serve cold.

It is indeed a delicate dainty as the poet describes it, but the syrup must not cook for as long a time as his recipe advised. My dessert was an excellent candy resembling a Chinese sweet long before it was time to add the yolks of eggs. So now it is made with 1 cup water and boiled to 220°.

This is a

CUSTARD JOSEPHINE BAKER

Beat 3 eggs with 3 tablespoons sugar. Mix 2 tablespoons flour in a little milk and add 2 cups more milk. Mix with sugar-and-egg mixture and strain. Add 2 teaspoons kirsch and 3 tablespoons liqueur Raspail. Add 3 bananas cut in thin slices and a few tiny pieces of the zest or rind of a lemon. Mix well, pour into a fireproof dish and cook in preheated 400° oven for 20 minutes. Serve cold.

Raspail is a liqueur for which it will probably be necessary to substitute another.

Rossini, who was inordinately fond of truffles, created this salad, but only after Alexandre Dumas, junior, had put before the public a better version of it.

ROSSINI'S SALAD

Four parts boiled sliced potatoes and one part sliced truffles that have been cooked in champagne and are served with the usual *vinaigrette* dressing of three parts olive oil, one part vinegar, salt and pepper.

And here is the proof that the master is greater than the follower:

ALEXANDRE DUMAS JUNIOR'S FRANÇILLON SALAD

This is the recipe as he gives it in his play *Françillon*, first produced at the Comédie Française.

ANNETTE: Cook some potatoes in *bouillon*, cut them in slices as for an ordinary salad, and while they are still warm season them with salt, pepper and a very good fruity olive oil, vinegar. . . .

HENRI: Tarragon?

ANNETTE: Orléans is better, but that is not of great importance. What is important is a half glass of white wine, Château Yquem if it is possible. A great deal of herbs finely chopped. At the same time, cook very large mussels in a *court-bouillon* with a stalk of celery; drain them well and add them to the potatoes.

HENRI: Less mussels than potatoes?

ANNETTE: A third less. One should taste the mussels little by little. One should not foretaste them, nor should they obtrude. When the salad is finished, lightly turned, cover it with round slices of truffles; a real *calotte* (or crown) for the connoisseur.

HENRI: And cooked in champagne?

ANNETTE: That goes without saying. All this, two hours before dinner, so that the salad is very cold when it is served.

HENRI: One could surround the salad with ice.

ANNETTE: No, no, no. It must not be roughly treated; it is very delicate and all its flavours need to be quietly combined.

It is a combination that is exquisitely and typically a French one. Why it is more popularly known as Japanese Salad no one has been able to tell me.

The proportions for it as made today are—2 lbs. potatoes, 3 quarts mussels, truffles—as many as the budget permits—3/4 cup olive oil, 4 tablespoons vinegar, 1 tablespoon herbs, 1/4 teaspoon peppers, little salt (the mussels are salty), 4 cups *bouillon* and 1 stalk of celery.

VIII.

FOOD IN THE UNITED STATES IN 1934 AND 1935

WHEN DURING THE SUMMER OF 1934 GERTRUDE STEIN COULD NOT decide whether she did or did not want to go to the United States, one of the things that troubled her was the question of the food she would be eating there. Would it be to her taste? A young man from the Bugey had lately returned from a brief visit to the United States and had reported that the food was more foreign to him than the people, their homes or the way they lived in them. He said the food was good but very strange indeed—tinned vegetable cocktails and tinned fruit salads, for example. Surely, said I, you weren't required to eat them. You could have substituted other dishes. Not, said he, when you were a guest.

At this time there was staying with us at Bilignin an American friend who said he would send us a menu from the restaurant of the hotel we would be staying at when Gertrude Stein lectured in his home town, which he did promptly on his return there. The variety of dishes was a pleasant surprise even if the tinned vegetable cocktails and fruit salads occupied a preponderant position. Consolingly, there were honey-dew melons, soft-shell crabs and prime roasts of beef. We would undertake the great adventure.

Crossing on the *Champlain* we had the best French food. It made me think of a college song popular in my youth, "Home Will Never Be Like This." If the food that awaited us at the Algonquin Hotel did not resemble the food on the French Line it was very good in its way, unrivalled T-steaks and soft-shell crabs and ineffable ice creams.

Mr. Alfred Harcourt, Gertrude Stein's editor, had asked us to spend Thanksgiving weekend with Mrs. Harcourt and himself in their Connecticut home, and there we ate for the first time, with suppressed excitement and curiosity, wild rice. It has never become a commonplace to me. Carl Van Vechten sends it to me. To the delight of my French friends I serve

WILD RICE SALAD

Steam 1/2 lb. wild rice.

1/2 lb. coarsely chopped mushrooms cooked for 10 minutes in 3 tablespoons oil and 2 tablespoons lemon juice, 2 hard-boiled eggs coarsely chopped, 1 green pepper finely chopped, 1 1/2 cups shelled shrimps, all lightly mixed and served with

AÏOLI OR AÏLLOLI SAUCE

Press into a mortar 4 cloves of garlic, add a pinch of salt, of white pepper and the yolk of an egg. With the pestle reduce these ingredients to an emulsion. Add the yolk of an egg. You may continue to make the sauce with the pestle or discard it for a wooden fork or a wooden spoon or a wire whisk. Real Provençal *Aïoli* makers use the pestle to the end. With whatever instrument you will have chosen you will commence to incorporate drop by drop an excellent olive oil. When the egg has absorbed about 3 tablespoons of the oil, add 1/2 tablespoon lemon juice. Continuing to stir, now add oil more briskly. When it soon becomes firm again add 1 dessertspoon tepid water (I repeat, tepid water). Continue to add oil, lemon juice and tepid water. The yolk of 1 egg will absorb 1 cup and 2 tablespoons oil, 1 1/2 tablespoons lemon juice and 2 dessertspoons tepid water.

Aïoli is, of course, nothing more than a garlic mayonnaise, a creamy mayonnaise. Mayonnaise with tepid water is creamier than without it. Mayonnaise should have more salt and pepper added to the yolk of egg than *Aïoli* as well as powdered mustard and paprika.

Gertrude Stein said she was not going to lunch or dine with anyone before lecturing, we would eat simply and alone. Before her first lecture she ordered for dinner oysters and honey-dew melon.

She said it would suit her. In travelling to a dozen states she deviated as little as possible from that first menu. Occasionally the oysters had to be replaced by fish or chicken. From the beginning the ubiquitous honey-dew melon bored me. Melons to me are a hot-weather refreshment. Rooms heated to 70° and over do not replace the sun. In any case, I prefer the flavour of Spanish melons to honey-dew and Persian melons. So the most fantastic dishes were experimented with, anything except what sounded like drug-store specialties.

Gertrude Stein continued with her satisfactory *régime* on the days of lectures. On the other days we fared more lavishly with friends in their homes and at restaurants, at first in New York, and then an excellent dinner at the inn at Princeton, at the Signet Club at Harvard with half a dozen of its members and no one else at Gertrude Stein's request, and very well at Smith College. Then we stayed with delightful people in an old historic house amidst rare and beautiful furniture and objects and dined and lunched with exquisite eighteenth-century porcelain, crystal and silver on a precious lace tablecloth, and left, quite starved, to find late in the afternoon fifty miles away an unpretentious but carefully cooked meal in a small town—oysters, roast turkey and its accompaniments and an unusually good rice pudding were not beyond our capacity. We asked to see the cook to thank her, and she gave me the recipe for

RICE PUDDING

Thoroughly wash 1/4 lb. rice, cook in double boiler in 1 quart milk with a pinch of salt. Stir the yolks of 8 eggs with a wooden spoon gradually adding 1 cup sugar and 5 tablespoons flour. Stir for 10 minutes and slowly add 2 cups scalded milk. Place over very low flame, stirring continuously until the mixture coats the spoon. Remove from heat and strain through a sieve, adding 1 teaspoon vanilla extract. When rice is quite tender, add slowly to egg-sugar-milk mixture. Then gently incorporate the beaten whites of 3 eggs.

Pour into buttered mould and cook in 350° oven for 20 minutes. Do not remove from mould until tepid. Serve with

VANILLA CREAM SAUCE

Stir the yolks of 6 eggs thoroughly with 1 cup sugar. Add 2 1/4 cups scalded milk. Stir over very low flame with wooden spoon until the mixture coats the spoon. Remove from flame and add 1 tablespoon best kirsch. Strain through hair sieve. Stir occasionally until cold enough to put into the refrigerator. Before serving gently add 1 cup whipped cream.

Gertrude Stein's and Virgil Thomson's opera was to be given in Chicago. She had never heard it, so when Bobsie Goodspeed telephoned that we ought to fly out there to hear it—there would not be time between lectures to go there by train—Gertrude Stein said she would but only under the protection of Carl Van Vechten. After a perfect performance of *Four Saints in Three Acts*, Bobsie gave a supper party. She was known to have a perfect *cuisine.* Of the many courses I only remember the first and the last, a clear turtle soup and a fantastic *pièce montée* of nougat and roses, cream and small coloured candles. The dessert reminded me of a postcard Virgil Thomson once sent us from the Côte d'Azur, delightfully situated within sight of the sea, pine woods, nightingales, all cooked in butter. This is the recipe for

CLEAR TURTLE SOUP

Soak 1/2 lb. sun-dried turtle meat in cold water for four days changing the water each day. On the fourth day prepare 1 stalk celery, 1 leek, 1 carrot, 2 onions and 1 turnip. Put 12 peppercorns, 3 cloves, 8 coriander seeds, a sprig of basil, of rosemary, of marjoram and of thyme in a muslin bag. Put the vegetables, the bag of spices and condiments and the turtle meat in a large stewpan. Cover with

4 quarts stock and bring to the boil uncovered, skim thoroughly, cover and simmer gently for 8 hours at least. It may be necessary to add more stock, in which case add very little at a time and be certain that it is boiling. When the turtle meat is quite tender, remove from pan and put aside. Strain the soup through muslin. When the fat rises to the surface, carefully remove all of it. To clarify the soup add the whites of 3 eggs and the juice of 1/2 lemon. Put over moderate heat and bring to the boil whisking continuously. When it boils, reduce heat, cover. In 10 minutes, strain through muslin. It will have come beautifully limpid. Cut the turtle meat into 1-inch slices, put into strained soup, add salt and a good pinch of cayenne, 1/2 cup best dry sherry per quart of soup. Serve hot. A tasty, nourishing but light soup.

We were driven through a winter landscape to a women's college where Gertrude Stein had accepted an invitation to dine with some members of the faculty. The dining-room was really a huge mess hall with acoustics that made a pandemonium of the thousands or was it only hundreds of voices. It was the beautiful young women students who were making this demoniacal noise. No wonder we had always thought of the graduates of the college as sirens, tragic and possibly damned. A restricted dinner was served in a manner appropriate to the surroundings. Gertrude Stein asked if she might have a soft-boiled egg and an orange.

After Gertrude Stein had lectured in New England, we went to Wisconsin, Ohio, Illinois and St. Louis, where the cooking was uniformly good with the exception of a superlative lunch given by a friend of Carl Van Vechten at her vast estate near Minneapolis. The drawing-rooms and dining-room were filled with flowers, largely orchids, the first Tiepolo blue ones we had ever seen. The dining-room table had a bowl of several varieties of hot-house grapes with thin tendrils and tender leaves—and the snow steadily falling outside. Our hostess was in the tradition of a Dumas *fils* heroine, though she was, I believe, the original of Carl Van Vechten's

Tattooed Countess. It is unnecessary to say that the menu was entirely a French one, and therefore a recipe of one of its courses has no place here. The temptation however is too great. This is the way to prepare

LOBSTER ARCHIDUC

Thoroughly wash a live lobster weighing not less than 3 lbs. Plunge into boiling water, allow to cool in liquid. Cut it down the middle and then across, take off the two claws, put aside the coral or eggs. In a deep pan melt over hot flame 4 tablespoons butter and 4 tablespoons oil. When it bubbles, put the six pieces of lobster, in their shell, into the pan. Heat thoroughly, turn with a wooden spoon until each piece is coated with the butter and oil. Then cover the pan and reduce the heat. Cook gently for 1/2 hour. Drain the lobster. Remove all meat from the shell and replace in the pan with the sauce. Replace over heat. Reheat slowly over low flame. Add the coral or eggs, 1/4 cup brandy, 1/2 cup best port wine and 2 tablespoons whisky. Season with salt and cayenne pepper. Cover and allow to boil for 5 minutes. Add 2 cups heavy cream. All to boil. Add the yolks of 2 eggs, heat thoroughly but do not allow to boil. Add the juice of 1/4 lemon and 5 tablespoons butter in very small pieces, turn gently until melted. Serve. This dish has an illusive flavour.

When we were at St. Paul to our surprise and delight there was a telephone message from Sherwood Anderson. He had heard we were in the neighbourhood. He proposed calling for us and driving us down to meet his wife—they were staying with her sister and their brother-in-law—which he did, through miles of ice and snow-drifts, to sweet people and a festival dinner. It was the happiest of meetings. Of all the delicacies served, it is strange to remark that it was the first time we tasted mint jelly.

In Columbus, Ohio, there was a small restaurant that served meals that would have been my pride if they had come to our table from our kitchen. The cooks were women and the owner was a woman and it was managed by women. The cooking was beyond compare, neither fluffy nor emasculated, as women's cooking can be, but succulent and savoury. Later, at Fort Worth, there was a similar restaurant to which Miss Ela Hockaday introduced us. We were to fly out to California and the restaurant packed us a box of food that was the best picnic lunch ever was. It would be a pleasure to be able to order something approaching it when taking a plane today. Has food on the American planes—not the transatlantic flights but on interior routes—improved? It has not in Europe, it is incredibly bad, even worse than on trains. Do they cook these meals in the locomotive and in the fuselage?

At Detroit there was a strange incident at the hotel which seemed sinister to us. The hardened European visitors became frightened. Gertrude Stein had the habit of an hour's walk after the evening meal, improperly spoken of as dinner. To calm her mind, she went off for a walk, but in a short time she returned quite agitated. Not far from the hotel, from the loudspeaker on a tower with a revolving searchlight, a warning was being repeated that no one was to move until a gunman was caught. A murder had just been committed. Suddenly Joseph Brewer's name flashed into my head. Had we not said we would stay with him if we were in his neighbourhood? He was the president of Olivet College. So we telephoned him and said we would like to be rescued. He said he would come to collect us and our bags, which he did the next morning with a large part of his faculty in several cars. It was an invigorating drive through snow and bitter cold sunshine to Lansing, where we had a carefully prepared lunch. For dessert we had an old-fashioned

BIRD'S-NEST PUDDING

Butter a porcelain pudding dish, slice 8 apples into it, sprinkle with sugar. Pour over them a batter made of 1 cup sour cream, 1 cup flour. Mix well, add the yolks of 3 eggs and 1 cup milk in which has been mixed 1 scant teaspoon baking soda. Beat the whites of 3 eggs, fold into mixture. Bake for 1/2 hour in medium oven. Brush the top with melted butter and sprinkle with sugar. Brown for 10 minutes. Serve with sweetened heavy cream. This is a pudding we should not neglect.

With a couple of days' rest with Joseph Brewer and the students at Olivet we forgot the horrors of Detroit and started off again. With Gertrude Stein's cousins in their home near Baltimore we enjoyed our first southern hospitality. We went to see Scott Fitzgerald in Baltimore who, with tea, offered us an endless variety of *canapés*, to remind us, he said, of Paris. In Washington southern hospitality continued. There was no disparity between the inspired negress cook and the enormous kitchen over which she presided. The hospitality was so continuous that there was never time to ask her for a recipe from her vast repertoire. She made the cakes, ices, punches and sandwiches for the parties, and the elaborate lunches and dinners that succeeded each other. No trouble at all, she said, when one has all the best material one needs. A dish, my father once said, can only have the flavour of what has gone into the making of it.

In New York we picked up Carl Van Vechten who was going to Richmond with us to introduce us to some of his friends there. On the way we stopped at Charlottesville where Gertrude Stein was to lecture at the University of Virginia, and where we lunched extremely well with some of the faculty, who pleased us with their divided allegiance to Edgar Allan Poe and Julien Green. At an epicurean dinner at Miss Ellen Glasgow's I was paralysed to find myself placed next to Mr. James Branch Cabell, but his cheery, Tell me, Miss Stein's writing is a joke, isn't it, put me completely at my ease so that we got on very well after that.

At William and Mary we lunched in state with the president at the Governor's house. On the road to Charleston we lunched at an old

Planter's Hotel copiously and succulently, for which the French have the nice word *plantureux.* We were asked to lunch at Strawberry. Was the exquisite food more seductive than the incredible water gardens, was the preparation of the menus at the Villa Margharita more exciting than the avenues of camellias? I have never been able to decide. Now they are all one. Changing planes at Atlanta, Gertrude Stein was delighted to see on a huge sign near the airport, Buy Your Meat and Wheat in Georgia.

In New Orleans we found Sherwood Anderson again and he took us to lunch at Antoine's and at a smaller restaurant which we preferred where we ate for the first time

OYSTERS ROCKEFELLER

Place oysters on the half shell in preheated deep dishes filled with sand (silver sand glistens prettily). Cover the oysters thickly with 1/4 chopped parsley, 1/4 finely chopped raw spinach, 1/8 finely chopped tarragon, 1/8 finely chopped chervil, 1/8 finely chopped basil and 1/8 finely chopped chives. Salt and pepper some fresh breadcrumbs, cover the herbs completely, dot with melted butter and put for 4 or 5 minutes in a preheated 450° oven. Serve piping hot.

This dish is an enormous success with French *gourmets.* It makes more friends for the United States than anything I know.

In New Orleans I walked down to the market every morning realising that I would have to live in the dream of it for the rest of my life. How with such perfection, variety and abundance of material could one not be inspired to creative cooking? We certainly do overdo not only the use of the word but the belief in its widespread existence. Can one be inspired by rows of prepared canned meals? Never. One must get nearer to creation to be able to create, even in the kitchen.

Before leaving Miss Henderson gave us two bottles of orange

wine, wine that was still being made in her home. It wasn't until some weeks later that we opened one of the bottles in Chicago and found the wine to be pure ambrosia.

In Chicago we stayed in Thornton Wilder's flat. He had said it would be convenient for Gertrude Stein as it was close to the university where she was to lecture. There was an extensive view from the little flat. It was very exciting, compact and comprehensive. The kitchen, though no larger than a dining-room table, permitted one, with its modern conveniences and marketing by telephone, to cook with the minimum of time and effort quite good meals. Those days are still my ideal of happy housekeeping. Once again we had lovely food with Bobsie Goodspeed, and at old-fashioned restaurants with friends and a delicious dinner with Thornton Wilder at a lakeside restaurant. We even had guests for meals at the flat. The meat or fowl delivered in waxed paper was deposited from the outside hall into the refrigerator, as were also the vegetables, cream, milk, butter and eggs.

On to Dallas where we went to stay with Miss Ela Hockaday at her Junior College. It was a fresh new world. Gertrude Stein became attached to the young students, to Miss Hockaday and the life in Miss Hockaday's home and on the campus. Miss Hockaday explained that all good Texas food was Virginian. Miss Hockaday's kitchen was the most beautiful one I have ever seen, all old coppers on the stove and on the walls, with a huge copper hood over the stove. Everything else was modern white enamel. The only recipe I carried away with me was for cornsticks, not knowing in my ignorance that a special iron was required in which to bake them. But when we sailed to go back to France in my stateroom one was waiting for me, a proof of Miss Hockaday's continuing attentiveness. It was my pride and delight in Paris where it was certainly unique. What did the Germans, when they took it in 1944, expect to do with it? And what are they doing with it now?

At the university at Austin the faculty asked some of the students to meet Gertrude Stein after the lecture. A very stiff punch was served, but when I was about to light a cigarette I was asked not to do so. Only men smoked.

Then we were off to God's own country. It was even more so than I remembered it. If there were more people and more houses, there were compensatingly more fields, more orchards, more vegetables and more gardens. A great part of the United States that we had seen had been new to me, it was a revelation of the beauty of our country, but California was unequalled. Sun and a fertile soil breed generosity and gentleness, amiability and appreciation. It was abundantly satisfying. In Pasadena amongst olive and orange groves we saw our first avocado trees and their fruit offered for sale stacked in great pyramids, almost as common as tomatoes would be later in the season. Driving north we heard that the desert wild flowers were in bloom so we took a day off to see them and the date palms. Through acres of orchards and artichokes, we made our way north to Monterey where happy days of my youth had been spent in an adobe house where my friend Señora B. had been born. The story was that General Sherman had courted her in the garden of her home, and before leaving Monterey had planted a rose tree later to be known as the Sherman rose. By the time I stayed with her she was an exquisite wee old lady with flashing black eyes. She would throw one of her shawls over my shoulders and say with a devilish glint, Go out and stand under the rose tree and let the tourists from Del Monte take your photograph. They will try to give you four bits but you may continue to turn your back on them. Señora B. made a simple Spanish sweet of which *Panoche* is the coarse Mexican version. She made it like this and unpretentiously called it

DULCE (1)

In a huge copper pan put quantities of granulated sugar, moisten with cream, turn constantly with a copper spoon until it is done. Then pour into glasses.

Señora B. said the longer it cooked the better the flavour would be. Señora B. would start it early in the morning and would entrust it to me when she went to mass. It was a compliment I could have dispensed with. As she was so little she stood on a footstool before her charcoal fire. In her simple but voluminous dark cashmere clothes she looked like a Zurbaran angel.

Here is my version of the

DULCE (2)

Put 2 cups sugar and 1 cup thin cream in a saucepan and bring to a boil. Then at once lower the heat and cook very slowly, stirring continuously for about an hour. It will become heavy and stiff and will have the colour of its flavour. There are people who like it a lot.

We had stopped at Monterey so that Gertrude Stein could see the house of Señora B. but it was no longer where it had been. A traffic policeman came up to us and asked us roughly what we were trying to do. To find Señora B.'s home, I said. It used to be here. That's right, he said, but years ago one of those rich easterners came out and bought it and carted it away into the hills. Carted an adobe house away, I muttered. But he wafted us on.

At Del Monte cooking was still of passionate interest to the management of the hotel. The same careful attention was given to the kitchen as to the vegetable and flower gardens. Grilled chicken and turkey broilers, spring lamb, cooked on a spit and basted by brushing it with a bunch of fresh mint, served with gooseberry jelly and an iced *soufflé*, were still unrivalled experiences.

This is the way to make

GOOSEBERRY JELLY

Take the tips and stalks off 6 lbs. gooseberries, put in a pan over a low flame with 4 pints water. Simmer until the berries are tender.

Turn into a jelly bag and let the juice run through. Weigh the juice. Place over high flame and boil briskly for 15 minutes. Add equal weight of sugar. Mix thoroughly and bring to the boil. Boil for 15 minutes or until it jellies.

Here is the recipe for the ineffable

ICED SOUFFLÉ

Put 2 cups sugar in heavy enamelled saucepan with 8 yolks of eggs and 1 whole egg over lowest flame. Beat with a rotary beater until it is quite thick. This will take some time. When it makes pointed peaks when the egg beater is removed, take from the stove and flavour with 1 tablespoon kirsch or anisette. Place on ice to cool. Pour into a *soufflé* dish and sprinkle on top 3 macaroons dried in the oven, rolled and strained. Put in the refrigerator for 3 hours. This is a particular favourite with men.

At Del Monte Lodge we ate for the first time abalone, and thought it a delicious food. It was served in a cream sauce in its shell, lightly browned with breadcrumbs without cheese, we gratefully noticed. Abalone has a delicate flavour of its own and requires no barbecue or barbarous adjuncts.

In San Francisco we indulged in gastronomic orgies—sand dabs *meunière*, rainbow trout in aspic, grilled soft-shell crabs, *paupiettes* of roast fillets of pork, eggs Rossini and *tarte Chambord*. The *tarte Chambord* had been a specialty of one of the three great French bakers before the San Francisco fire. To my surprise in Paris no one had ever heard of it.

At Fisherman's Wharf we waited for two enormous crabs to be cooked in a cauldron on the side-walk, and they were still quite warm when we ate them at lunch in Napa County. Gertrude Atherton took us to lunch at a restaurant where the menu consisted entirely of the most perfectly cooked shell-fish, to her club where the cooking was incredibly good, and to dinner at a club of writers where conversation excelled.

And then the dearest friend sent us a basket of fruit and flowers, fit subject for an Italian painter of the Renaissance, and we tasted for the first time passion fruit. We had known passion-fruit syrup in Paris and thought its flavour exquisite (it made a wonderful ice cream). And now we were told that passion fruit was the fruit of the passion-flower vine. Surely not from the passion-flower vine that has climbed a wall in every garden I ever had.

Then the time had come when we would have to leave California, to leave the United States, to go back to France and cultivate our garden in the Ain. Above everything else I enjoyed working in that garden, but leaving the United States was distressful.

It was not until we were on the *Champlain* again that I realised that the seven months we had spent in the United States had been an experience and adventure which nothing that might follow would ever equal.

IX.

LITTLE-KNOWN FRENCH DISHES SUITABLE FOR AMERICAN AND BRITISH KITCHENS

THESE DISHES SHOULD ADD VARIETY TO AMERICAN AND BRITISH MENUS. In France they are no longer novelties nor creations, nor have they the distinction of being distinctive, which, as defined by a cousin of Gertrude Stein, is something that is done six weeks before all the world is doing it. On the contrary, they are most of them a slow evolution in a new direction, which is the way great art is created—that is, everything about is ready for it, and one person having the vision does it, discarding what he finds unnecessary in the past. Even a way of cooking an egg can be arrived at in this way. Then that way becomes a classical way. It is a pleasure for us, perhaps for the egg.

It is, of course, understood that there are always those who rush in and irreverently add a dash or a pinch from a bottle, a tin or a package and feel that some needed flavour has been found. This, a matter too literally of taste, is not arguable. It is a pleasure to retire before such a fact.

For the preparation of these dishes, certain standard sauces and what the French call composed butters are necessary. Those used in these recipes follow these standards and conventions. To commence then at the beginning:

I

COLD HORS D'OEUVRE

MUSSELS

After thoroughly scrubbing and rinsing, put 2 quarts mussels in a saucepan over highest heat with 3/4 cup dry white wine,

1 tablespoon crushed shallot, 2 stalks of parsley, 1 twig of thyme, 1/4 laurel leaf, salt and pepper. Cover. About 2 minutes after it has come to a boil again look to see if the shells have opened. As soon as they have remove from flame, drain thoroughly, remove from shell, cool, chill. Just before serving mix with 3/4 cup Tartar sauce. Serve very cold.*

EGG PLANT À LA PROVENÇALE

Wipe but do not peel 6 *aubergines* or egg plants, cut in slices of 1 1/2 inches, salt and pepper; put them in saucepan over medium heat in which 6 tablespoons olive oil are bubbling. Brown lightly on all sides, remove from flame, drain the *aubergines.* In the oil in which they have been cooked, put 3 blanched and skinned tomatoes that have been coarsely chopped. Add salt and pepper, 1 clove of crushed garlic and 2 diced anchovies. Cook over low heat until tomatoes are cooked. Remove from flame and add 1 tablespoon chopped basil. Place 1 teaspoon of this sauce on each slice of egg plant. On the sauce place a thin slice of lemon, on the lemon place a rolled anchovy, and in the anchovy stand 1 black olive upright.

SMALL FISH IN THE ORIENTAL MANNER

For 4 fish—any small fish but preferably red mullets weighing 1/2 lb. each—clean, dry and remove scales if necessary. Place in hot olive oil in a frying pan over great heat. Fry on both sides for 3 minutes. Remove from heat. Put in well-oiled fireproof dish and cover with 8 tablespoons *purée* of tomato that has been previously cooked for 10 minutes over low heat with 1/3 cup white wine, 1/4 teaspoon powdered saffron, 1/4 teaspoon powdered thyme, 1/4 teaspoon powdered laurel and 1/8 teaspoon powdered or

* *Note.* The bay laurel *(Laurus Nobilis)* is the one to use here and wherever else a laurel leaf is needed in my recipes.

ground coriander seed, 1 clove of crushed garlic, 1 teaspoon chopped parsley and salt and pepper. At the last moment add 4 tablespoons olive oil but without allowing it to boil. Cover and cook in preheated 400° oven for 8 minutes. Serve chilled.

STUFFED CUCUMBERS

Cut unpeeled cucumbers in half lengthwise. Boil for 2 minutes. Remove from heat. Put under cold-water tap, drain and dry thoroughly. When cold, with a sharp knife hollow out within 1/4 inch of the skin and fill with previously cooked, chilled and diced string beans and green peas. Chill thoroughly and cover with a green mayonnaise. Sprinkle with minutely cut chives.

GREEN MAYONNAISE

Put the yolk of 1 egg in a bowl with 1/2 teaspoon salt and 1/4 teaspoon pepper. Stir well. Add drop by drop 2 tablespoons olive oil, constantly stirring. When it commences to stiffen add a few drops lemon juice and pour the oil in more quickly. It will require 3/4 cup of oil and the juice of 1/2 lemon. This mayonnaise must be particularly firm. Then gradually add the following *purée*:

Take equal parts of leaves of cress, spinach, chervil and tarragon, boil them in unsalted water for 2 minutes, drain and put under cold-water tap and press out the water. Pound in a mortar until they are reduced to a pulp that can be strained through a fine muslin. They should make 1/4 cup. Add to the mayonnaise, to which the greens will give not only a colour but a flavour.

FROGS' LEGS À LA PARISIENNE

In a saucepan over medium heat place the frogs' legs covered with dry white wine, salt and pepper, the juice of 1 lemon (for 50 frogs' legs). Poach for 10 minutes. Remove from heat and drain. In a bowl or dish place an equal volume of diced potatoes salted and peppered. Mix with mayonnaise (1/4 cup mayonnaise for 2 cups potatoes). Place the frogs' legs on the potato salad. Cover with mayonnaise.

CAULIFLOWER WITH MUSTARD CREAM SAUCE

Boil the flowerets of cauliflower in salted water for 10 minutes. Drain, cool and chill. Before serving, pour over them this sauce.

In a saucepan melt 1 tablespoon butter. Add 1 tablespoon flour and 1 tablespoon mustard. Mix thoroughly, add 2/3 cup boiling water. Mix with a whisk. Add 2 yolks of eggs mixed with 1 tablespoon whipped cream. Beat before pouring over flowerets and serve at once.

II

HOT HORS D'OEUVRE

POACHED EGGS BABOUCHE

For four, cut 2 large tomatoes in half, remove seeds and juice and grill. Place 1 tablespoon boiled rice in each half tomato, on which place a poached egg. Cover with 2 cups cream sauce, in which 1 teaspoon curry has been mixed and the sauce brought to a boil.

FRIED CREAM WITH CHEESE

Stir 2 whole eggs and 3 yolks, add gradually 1 cup and 2 tablespoons flour, stir carefully. The mixture should be perfectly smooth. Add 1/4 teaspoon salt, a pinch of pepper and nutmeg. Add 1 quart hot milk. Place over low heat and stir until the mixture is very thick, but do not allow to boil. Remove from heat and add 1 cup grated Swiss or Parmesan cheese. Spread on buttered marble, or on a large dish. The mixture should be about 1 inch thick. See that the edges are even. When cold, cut into squares. Dip into beaten egg, then into dried fine breadcrumbs. At the last moment, before serving, fry in deep oil.

OYSTERS SAUCE MORNAY

Poach oysters to heat, drain thoroughly and replace in shells. *Sauce Mornay* is a cream sauce with cheese in this proportion: If the oysters are large, 5 dozen will require 2 cups thick cream sauce to which has been added while still over low flame 2 tablespoons grated Swiss cheese and 2 tablespoons grated Parmesan cheese. Add in small pieces 4 tablespoons butter. Do not allow to boil. Remove from heat, completely cover the oysters. Sprinkle with a little grated cheese and place in 450° oven for 4 or 5 minutes. Serve at once.

OMELETTE PALERMITAINE

Mix 8 whole eggs with 4 tablespoons diced truffles, 3 tablespoons grated Parmesan cheese and 2 tablespoons lightly browned butter. With a fork stir until the yolks and whites are thoroughly mixed. In a very hot frying pan melt 3 tablespoons butter, allowing it to become very slightly browned. Pour the eggs into the frying pan. In 1/2 minute bring the edges to the centre with a fork so that the eggs that are liquid may be cooked. If necessary tip the frying pan. In 2 minutes more the omelette will be done. Place 3 tablespoons of hot *purée* of tomatoes, to which has been added 2 tablespoons butter, on one side of the omelette. Fold the other side of the omelette over across the centre, and place on a hot serving plate. Surround with a thick tomato sauce.

FRIED OYSTERS

Poach the oysters for 1 minute in their own water. Remove from heat, drain and dry thoroughly. Sprinkle generously with olive oil, salt and pepper. Marinate for 1 hour. Cover each oyster with salted frying batter. Fry in deep fat, remove as soon as brown. Place on hot serving dish in a mound, and surround with very thin slices of lemon. In the centre of the mound place a quantity of slightly fried parsley.

SOUP

MIMOSA SOUP

Boil in salted water 1 cup string beans. When tender, depending upon size and freshness, remove from heat, drain and cut into cubes. Mash through a very coarse sieve the yolks of 4 hard-boiled eggs. Just before serving, bring to a boil 1 quart *bouillon.* Add diced string beans and the yolks of eggs. Bring to a boil again, remove at once and serve.

CREAM OF GREEN-PEA SOUP

Put 2 cups green peas in 6 cups boiling water in a saucepan over medium heat. Add 1 medium-sized onion, salt, pepper and 1 stalk of parsley. Boil uncovered until peas are tender—about 25 minutes, depending on age and freshness. Prepare small *croûtons.* Chop fine 1 tablespoon chervil. Mix 3 yolks of eggs with 6 tablespoons cream, add chervil. Before serving, remove onion and parsley from water in which peas have boiled. Add, at once, 1/2 cup butter cut in small pieces, do not stir but tip the saucepan in all directions. Pour over cream and eggs. Add *croûtons* and serve.

CONSOMMÉ WITH PARMESAN CHEESE CROÛTONS

Mix 4 tablespoons grated Parmesan cheese with 1 tablespoon flour. Separate the yolks from the whites of 2 eggs. Beat the whites of the eggs, fold in the yolks, the grated cheese and flour, and salt and pepper. Spread this mixture on slightly buttered bread to a thickness of 1/3 inch and cook in medium oven for 20 minutes. While still hot cut into squares, rounds or ovals, and serve at the same time as 6 cups hot *bouillon.*

ONION SOUP

Brown 4 thinly sliced onions in 2 tablespoons butter in a saucepan over low heat stirring with a wooden spoon. In 20 minutes, sprinkle 1 teaspoon sugar over the onions. Add 6 cups boiling water, salt and pepper. Boil, covered, for 10 minutes. In a fireproof dish or casserole place 1/2 lb. bread cut in slices of 1/3-inch thickness. Cover each piece of bread with very thin slices of Swiss cheese. It will take a little more than 1/4 lb. cheese. Place them in the soup dish and pour over them the contents of the saucepan. The slices of bread will rise to the surface. Sprinkle on the top 4 tablespoons melted butter. Put the casserole in 375° preheated oven for 20 minutes. Serve very hot.

FISH

MEURETTE

This is to Burgundy what the Bouillabaisse is to Marseilles. The fish are mixed river fish and should be small—trout, perch, eel, pike and carp are most suitable. Use all five of them if possible. Clean and remove scales and fins, and wash 3 lbs. fish.

If any of them are too large for one serving, cut in slices. Place in a saucepan over low heat 1 bottle red wine most nearly resembling Burgundy wine, with 1 large onion, 3 medium-sized carrots, the white ends of 2 leeks, 3 cloves of garlic, salt and pepper, a very small piece of nutmeg or a pinch of powdered nutmeg, a twig of thyme, a laurel leaf and 1 clove. Simmer for 1/4 hour. Strain this over the fish that have been placed in fireproof casserole. Add1/4 cup brandy, cover and cook for 1/2 hour. Add 5 tablespoons butter cut in small pieces. Do not stir but tip casserole in all directions. Do not allow to boil. At the bottom of the serving dish place 8 pieces of bread toasted, and rubbed with garlic. Cover with the fish and pour the *bouillon* over it.

BAKED SHAD

Remove scales and fins of a 3 lb. shad. Wash thoroughly but do not allow to remain in water. Drain and dry. Cut in slices. In a generously buttered fireproof dish, place a thin sprinkling of 4 finely chopped shallots, 1/4 lb. finely chopped mushrooms and 1 tablespoon finely chopped parsley. Place the sliced fish on this, reconstructing the fish, head at one end, tail at the other. Add 1 cup dry white wine. Sprinkle 2 tablespoons melted butter on the fish. Cook in preheated 370° oven, baste frequently. After 20 minutes add 1/2 cup cream and cook for 10 minutes longer, having increased heat to 425°.

FILLETS OF SOLE PERINETTE

Salt and pepper 4 large fillets of sole. Prepare 2 cups mashed potatoes thoroughly mixed with 2 tablespoons cream, 1 egg, salt and pepper. Mash through fine sieve 2 yolks hard-boiled eggs. Chop very fine 1/4 cup truffles, mix yolks of eggs and truffles with potatoes. Spread this mixture on two of the fillets, cover each with another fillet, press them together. Cover them with a beaten egg mixed with 1 tablespoon olive oil and 1 tablespoon water. Cover with grated breadcrumbs. Melt in frying pan over medium heat 1/2 cup butter. Brown the fillets on both sides in this, reducing the heat so that the fillets will be cooked through. In a preheated serving dish spread 2 cups mashed potatoes, to which has been added 4 tablespoons browned butter. Place the fillets on the potatoes, and on the fillets place thin slices of truffles alternating with thin slices of lemon.

FRESH COD MONT-BRY

Have 3 lbs. fresh cod cut in thin slices. Heat 3/4 cup olive oil in frying pan over medium heat. Place the slices of fish which have been rubbed with salt and curry powder in the hot oil. Reduce heat and brown fish on both sides. Place them on serving dish on which has been spread a tomato and egg plant *purée* which has been prepared by cooking in a covered saucepan 6 peeled and chopped tomatoes and 4 peeled and chopped egg plants in 3 tablespoons olive oil with 4 cups dry white wine for 20 minutes. Salt and pepper, and strain. On each slice of fish place a thin slice of lemon and on the lemon place a tablespoon of diced pimiento heated in olive oil. Serve with

Curry Sauce

Chop 1 large onion and cook it in covered saucepan over low heat in 3 tablespoons olive oil for 20 minutes. Add 1 cup *Béchamel* sauce. Strain and replace over low heat. Add a pinch of saffron and 2 tablespoons curry powder, salt and pepper. Bring to a boil and serve in sauce boat at the same time as the fish.*

* *Note. Béchamel* is a basic white sauce made with equal quantities of flour and butter, cooked in enough milk to make a creamy mixture.

DEVILLED SMELTS

Clean, remove fins, wash and dry 6 smelts. In a bowl mix 1 1/2 tablespoons powdered mustard, 1 1/2 tablespoons French mustard, 3 tablespoons olive oil, 3 yolks of eggs, 1/2 teaspoon salt, 1 tablespoon anchovy paste and 1/4 teaspoon cayenne pepper. Mix thoroughly and spread on both sides of each fish. Pour melted butter over the fish and cover them with fine cracker crumbs. Grill over low heat. Serve with

Devil Sauce

Put 1/4 cup dry white wine, 1 tablespoon vinegar, 1 tablespoon chopped shallots and 1/4 teaspoon pepper in a small saucepan over low flame. Simmer, stirring constantly with wooden spoon until reduced to half its volume. Add 1 cup tomato juice and 1 cup white wine. Bring to a boil, simmer and add 1 1/4 tablespoons flour thoroughly mixed with 1 1/4 tablespoons butter. Stir and watch carefully. The sauce should be perfectly smooth. Mix 1 tablespoon dry mustard, 1 tablespoon French mustard and 1/2 tablespoon anchovy paste, 1/4 teaspoon cayenne pepper, 1/4 teaspoon powdered saffron. Stir carefully, and simmer for 10 minutes. Remove from flame, strain into sauce boat and serve at the same time as the fish.

STEWED MACKEREL WITH PAPRIKA

If the mackerel are small take 4; if large, 4 slices. Clean, remove fins, wash and dry. In a stew pan melt 6 tablespoons butter over low heat, add 4 medium-sized onions cut in fine slices. Cover and simmer, shaking the pan from time to time, for about 20 minutes or until the onions are transparent. Then add 1 teaspoon paprika, salt and pepper. Place the fish on the onions with salt and pepper. Add 1 cup hot water. Cover and slightly increase heat. Cook for 20 minutes. Add 1 cup thinly sliced mushrooms.

Cover and continue to cook for 8 minutes more. With a perforated spoon remove fish to hot serving dish. Add to saucepan a sauce prepared in advance of 2 tablespoons butter mixed with 1 1/2 tablespoons flour. Place over low heat, gradually add very hot water, stir and cook for 5 minutes. Add salt, pepper and 1/2 teaspoon paprika. Cook for 5 minutes more. Then add 3/4 cup heavy cream. Add to contents of stew pan. Heat thoroughly but do not boil. Pour over fish and serve.

SALT CODFISH À LA MONÉGASQUE

Soak overnight 6 fillets salt codfish. In the morning put under cold-water tap. Drain, place in a bowl, cover with milk and soak for 2 hours. Then poach in water until tender. Drain, dry thoroughly, and cook over low flame in 1/2 cup olive oil in covered frying pan for 20 minutes. Shake the pan frequently and turn the fillets once. Place them on heated serving dish surrounded by small pieces of bread 1-inch thick lightly browned in oil. Serve with 1 1/2 cups thick tomato sauce in which 1/2 cup capers have been cooked.

RAY WITH BLACK BUTTER

The thorn-back ray is the best variety. Nothing is removed from the large ones, as they are sold ready to cook. The small ones require to be scraped, the gall, head and tail removed carefully, the fish cleaned and washed. Put 3 lbs. ray in a saucepan with cold water to cover, with 1 tablespoon salt and 2 tablespoons vinegar. Bring to a boil over medium heat. When the water boils, reduce to the least possible heat and allow 1/2 hour for a large piece of ray and 20 minutes for small ones. Remove from water, drain thoroughly. Place wrapped in a cloth in tepid oven with door closed while the black butter is being prepared.

Place 1 cup butter in frying pan over medium heat with 1 teaspoon salt, 1/4 teaspoon pepper and 1 cup unchopped parsley. Add juice of 1 lemon. Pour butter and parsley at once over fish, which has been placed on serving dish, and serve.

POULTRY

CHICKEN IN HALF MOURNING

Cook 4 truffles, in Madeira to cover, over lowest flame in covered saucepan for 20 minutes. Remove from flame and allow to cool in wine. Place a fine chicken in a *cocotte* with a tight-fitting cover. Pour over it to half its height hot veal *bouillon* with 1 teaspoon salt, 1/2 teaspoon pepper, a bouquet of parsley and thyme and 1/2 laurel leaf. Bring to a boil over medium heat, then gradually reduce to lowest flame. One hour will cook the chicken. For the sauce, heat in saucepan over medium heat 1/2 cup butter, add 3 tablespoons flour, gradually add 3 cups veal *bouillon* and 1/4 cup chopped mushrooms, stirring with a wooden spoon. As soon as it comes to a boil reduce heat and simmer uncovered for 1/2 hour. Stir to prevent scorching or burning. Remove from heat, strain, skim, add 1/2 cup heavy cream and 1 1/2 finely chopped truffles. Cut chicken in serving pieces, and place on heated serving dish. Heat the sauce but do not allow to boil. Pour over chicken, and on each piece place a slice of truffle and serve.

BRAISED CHICKEN STUFFED WITH NOODLES

Poach for 6 minutes in boiling salted water 1 1/2 cups noodles cut in narrow strips. Remove from flame, drain, place under cold-water tap and drain again. Place in a bowl, and, mixing lightly with a fork, add to them 1/2 cup grated Parmesan, 1/2 cup grated Swiss cheese and 3/4 cup heavy cream. Add 3/4 cup small mushroom caps, salt and pepper. Mix well. Stuff the chicken with this dressing. Skewer or sew the opening together. Skewer legs and wings to keep in place during cooking. Put 4 tablespoons butter in enamel-lined pot over medium flame. Brown the chicken lightly in the butter. Add 1 cup hot chicken *bouillon*. Cover the pot and lower the flame. Simmer for 1 hour, basting, from time to time. Prepare a *Sauce Mornay*, which is a *Béchamel* to which grated cheese has been added. For this chicken the proportions are 2 1/2 tablespoons butter, 2 tablespoons flour, 2 cups milk, 1/4 teaspoon salt. Simmer for 1/2 hour. Stir with a wooden spoon frequently to prevent burning. Remove from heat, add 1/2 cup grated Parmesan cheese. Place the chicken in a fireproof serving dish. Cover with the *Sauce Mornay* and place for 10 minutes in 450° preheated oven to glaze.

COCK IN WINE

Cut a young cock or a young chicken in serving pieces. In an enamel-lined pot melt 3 tablespoons butter, add 3/4 cup diced side fat of pork, 6 small onions, 4 shallots and 1 medium-sized carrot cut in thin slices. Brown these in butter. Remove and place pieces of chicken in pot and brown over high heat. Add salt, pepper and 2 cloves of crushed garlic. Remove the browned pork fat, onions, shallots and carrot. Heat 3 tablespoons brandy, light and pour into pot. Sprinkle 3/4 tablespoon flour into the pot. Stir with a wooden spoon for 2 or 3 minutes, then add 1 cup fresh mushrooms and 1 cup hot good dry white wine. Increase heat, add pork fat and vegetables. Cook uncovered for 1/4 hour. Serve very hot.

A FINE FAT PULLET

A fine fat pullet is cleaned, the breast bone is removed with a small sharp knife and the bird is well skewered. Place 3/4 cup lard, 3/4 cup butter and 4 tablespoons olive oil in a casserole. Place the chicken in the casserole over medium heat. When the fats commence to boil reduce heat—the chicken should not brown. Add 2 tablespoons finely chopped shallots, 6 tablespoons finely chopped mushrooms, 2 tablespoons finely chopped parsley, 1/2 teaspoon salt, 1/4 teaspoon pepper. Cook for 10 minutes. Remove chicken and cover with contents of casserole. As the chicken cools, again cover with the fats and chopped vegetables. It may be necessary to do this several times. When the chicken is cold, wrap it in a thin slice of fat back of pork. Take six sheets of white paper large enough to cover the chicken completely and oil each sheet generously. Place the chicken on the first sheet and bring the end of the sheet to the middle of the breast of chicken. Continue with the other sheets. Tie the sheets so that the chicken is hermetically closed. Return chicken to casserole and cook for 1 hour in preheated 350° oven. To serve, cut the string of the sheets. If the two outside sheets are too dark from the oven heat, remove them but do not disturb the others. Send to the table wrapped in the remaining sheets.

GODMOTHER'S CHICKEN

In the cavity of a fine chicken put 1 cup chicken *bouillon*, 1/2 cup butter and a large bunch of tarragon which have been well mixed together, sew up the cavity and skewer the legs and wings. Place in a pot over medium heat, with 2 cups chicken *bouillon*, salt and pepper. When it comes to a boil, cover hermetically, reduce to lowest flame and simmer for 3/4 hour. Remove the chicken broth and the juice from the cavity of the chicken into the pot. Put the chicken aside and reduce the contents of the pot by half. Remove from flame and add 1/2 cup soft butter. Stir until melted, add salt, pepper, a pinch of powdered nutmeg and a pinch of cayenne pepper. Return to low heat. When very hot remove from heat and add the juice of 1/2 lemon and 2 tablespoons Madeira. Cover the chicken with the sauce.

CHICKEN MONTSOURIS

Cut a chicken down the back and flatten it with a mallet. Put it in a casserole with pepper, salt and 1/4 lb. butter over high flame. As soon as the butter bubbles put in preheated 400° oven. Baste frequently. After 35 minutes remove from oven. Put chicken on serving dish. Put the casserole over medium flame and add 2 finely chopped shallots. Heat 1/3 cup brandy, light it and pour into casserole. Add 2 tablespoons heavy cream, 1 teaspoon finely chopped tarragon. Stir with wooden spoon. Before it boils, add 5 tablespoons butter cut in small pieces. Tip casserole in all directions until butter is melted. Pour over chicken and serve.

MINUTE SQUABS (OR YOUNG PIGEONS)

Cut the squabs in half. Flatten them gently with a mallet. Paint with melted butter. In a frying pan over hot flame put for 2 squabs 1/2 cup butter. Keep turning the pieces so that they do not burn. They will be cooked in 12 minutes. When they are half cooked add 1 tablespoon onion juice. Place the squabs on preheated serving dish. Add to contents of pan 1 tablespoon cognac. Scrape pan to detach glaze. Stir in 1/2 cup hot *bouillon.* Add 1 teaspoon chopped parsley and pour over squabs.

ROAST SQUAB ON CANAPÉS

Place in the cavity salt and pepper. Cover squabs with thin slices of back fat of pork and tie securely. Put them in preheated 400° oven. Roast for 20 minutes. Baste twice. Cut the string and remove pork fat and place on *canapés.* The *canapés* are slices of bread dipped in melted butter and heated in a frying pan. They should not brown. Spread on them a *purée* of mushrooms and truffles. For 4 *canapés*, chop 1/2 cup mushrooms, pound them with a

potato masher through a sieve. Put them in a saucepan with 2 tablespoons butter and over medium heat cook uncovered until all the moisture has evaporated. Add salt and pepper, 1/4 cup *Béchamel* sauce and 1 tablespoon heavy cream. Simmer for 10 minutes or until quite thick. Put aside. Wrap to cover completely 4 truffles in fat back of pork. Tie each truffle securely. Place in casserole with 2 tablespoons Madeira. Cook covered over low flame for 20 minutes. Remove pork fat. Pound truffles through a sieve and add to mushroom *purée.* Add the juice in which the truffles have cooked to the pan in which the pigeons have roasted. Over medium flame add 2 tablespoons Madeira and 1/2 cup *bouillon.* Bring to a boil and pour over pigeons.

DUCK WITH BORDELAISE SAUCE

A 3 lb. duck will take about an hour in a preheated 400° oven. Put salt and pepper in the cavity. Paint the bird with 4 tablespoons butter and place in oven. Commence to baste after 1/4 hour. As the legs of a duck are proportionately heavier than those of a chicken, turn the duck on each of its sides for 10 minutes after it has been in the oven 1/2 hour.

Bordelaise Sauce

Press the liver of the duck and an extra one through a sieve and put aside. Put 2 finely chopped shallots with 2 tablespoons butter in saucepan over low heat. When all the moisture has evaporated, let them cook a few minutes longer to form a glaze. Add 1/2 cup *bouillon* in which 1 tablespoon tomato *purée* has been dissolved and to which 1 tablespoon flour and 1 tablespoon butter has been added. Add 4 tablespoons lighted brandy. Bring to a slow boil, add salt, pepper and a pinch of cayenne. Simmer for 10 minutes. Add 1/2 cup *bouillon.* Bring to a boil. Slowly pour half the contents of the saucepan over the two strained livers. Pour into the saucepan and replace over low heat, stirring constantly, but do not allow to boil. Add 4 tablespoons butter. Tip saucepan in all directions, do not stir. When butter is melted, pour into preheated sauce boat and serve at the same time as the duck.

YOUNG TURKEY WITH TRUFFLES

For a turkey weighing 6 to 7 lbs. you will require 3 lbs. of the fat that surrounds pork kidneys. Melt it and strain. Pour 4 cups of whole truffles into a quarter of the melted pork fat. The truffles should be cut in half or in quarters, according to size. Poach for 1/4 hour, mix with the rest of the fat and cool. When cool, fill the cavity of the turkey into which salt, pepper and a little nutmeg have been rubbed with the cold fat and truffles. Sew the cavity together, skewer wings and legs. Put aside in a cool spot—but not in the refrigerator—for two days in winter and one day in summer. When ready to roast, wrap the turkey in buttered paper. Roast in

400° preheated oven, allowing 10 minutes to the pound. Half an hour before it is ready to be taken from the oven, remove paper. Baste three times before removing from the oven. Place on serving dish. Skim the fat from the juice, add 3 cups of the *bouillon* of the giblets, neck and first joints of wings (that have been poached for 2 hours in water to cover, and then strained together with 1 bay leaf, a sprig of thyme, parsley and salt and pepper). Mix well to detach the glaze from the pan. Serve in sauce boat.

MEATS

SPRING NAVARIN

Ask the butcher to cut in square pieces—two to a serving—2 lbs. shoulder of mutton. In an iron or enamelled pot melt 2 tablespoons butter over high flame. When the butter is very hot, place the pieces of meat in it. Stir with a wooden spoon until all sides are lightly browned. Add salt, pepper and 2 tablespoons flour after the fat has been skimmed off. Stir for 10 minutes. Cover with hot water stirring constantly. The sauce should be perfectly smooth. Add a bouquet of thyme, parsley and a laurel leaf, a pinch of powdered nutmeg, a large onion with 1 clove stuck in it and a clove of crushed garlic. Add 1 lb. tomatoes, blanched, peeled, seeds removed and cut in small pieces. Bring to a boil, cover pot, and reduce heat. Wash 1 dozen new potatoes, dry and rub off the skins, scrape 1 dozen young carrots, wash and dry them and remove outer skin of 1 dozen spring onions, peel 1 cup young turnips and cut in quarters. Heat 4 tablespoons butter in a saucepan over medium heat. When hot, put in the turnips and sprinkle over them 1 teaspoon sugar. Brown them on all sides, stirring constantly. Then strain and put aside. Shell 1 cup green peas. Snap off the ends of 1 cup of very young string beans. After 1 hour's cooking remove bouquet and onion from pot. If there is any fat on the sauce, skim. Bring to a good boil and add the potatoes, carrots, turnips and onions. Cover and reduce to medium heat. Boil for 10 minutes in salted water, in separate saucepan, the green peas and beans. After the first vegetables have cooked in the pot for 1/2 hour, add peas and beans. Cook for 1/2 hour longer, skim the sauce again, and serve very hot in preheated hollow dish.

BOEUF BOURGUIGNON (2)

Ask the butcher to cut 2 lbs. upper shoulder of beef in square pieces, two pieces for one serving. Marinate in 6 tablespoons brandy for 6 hours, turning the meat frequently. In an iron pot over high flame put 4 tablespoons butter. While it is very hot place the meat in it in a single layer—no piece should cover another. Brown them on all sides, remove and put aside. Put into the pot 1 cup diced side fat of pork, brown and remove. Put into pot 12 small onions, brown and remove. From the pot remove about two-thirds of the fat. Put into the pot 1 tablespoon and 1 teaspoon flour. Stir with wooden spoon for 10 minutes and add the meat. Pour over it 3 cups hot red Burgundy wine or a wine that resembles it as nearly as possible and 1 cup hot water. Add a little salt, pepper, the diced pork fat, a bouquet of thyme, parsley and laurel leaf, a pinch of ground nutmeg and 1 crushed clove of garlic. Cover and when the contents of pot commence to boil reduce to medium heat. Cook for 2 1/2 to 3 hours. Add the onions and cook for 1/2 hour longer. Remove bouquet, if necessary skim, place on hollow serving dish and sprinkle over the meat 1 tablespoon finely chopped parsley.

It very much adds to the flavour of the dish to add at the last moment 2 cups small fresh mushrooms that have been cooked in a saucepan over low flame in 4 tablespoons butter for 10 minutes.

VEAL MARENGO

Ask the butcher to cut in pieces suitable for serving 2 lbs. shoulder of veal and remove bones. Put in iron pot over high flame 1/2 cup olive oil. When it commences to smoke put in the pieces of meat side by side. Brown on all sides and remove. Dice 1 cup mushrooms, place in pot, brown and remove them. Put 12 small onions in the pot, brown them and remove. Remove two-thirds of the oil from the pot and put in pieces of meat. With a wooden spoon mix 1 tablespoon and 1 teaspoon flour with the fat in the pot, brown for 2 or 3 minutes. Add 1 1/2 cups hot dry white wine and 1 1/2 cups hot water. Add 2 medium-sized tomatoes skinned, seeds removed and cut in cubes of about 1 inch, a bouquet of thyme, parsley and 1 laurel leaf, 2 crushed shallots and 1 clove of garlic, salt and pepper. Add to the pot the onions and mushrooms. Reduce heat, cover the pot, cook for 1 hour. Skim the oil, remove the bouquet, cover pot and cook for 10 minutes longer. Pour into preheated dish and surround with triangular pieces of bread, about 1/2 inch thick, that have been very lightly fried in olive oil.

FILLET OF BEEF ADRIENNE

Prepare a marinade with 1 bottle dry white wine—the better the wine the better the sauce—1/2 cup olive oil, a bouquet of thyme, parsley and 1 laurel leaf, 2 sliced medium-sized carrots, 1 sliced medium-sized onion, pepper and salt. In this, place a 3 lb. fillet of beef. Cover and put aside, but not in the refrigerator, for 48 hours. Turn the meat in the marinade twice a day. At the beginning of the third day remove meat from marinade and dry thoroughly. Wrap it in a thin slice of back fat of pork and tie this securely round the fillet. Place in a preheated oven 350°. Put the marinade, including the vegetables, in a saucepan over low flame. Turn the fillet after 1/4 hour. After 1/2 hour commence to baste the roast with 4 tablespoons of the hot marinade. Baste every 10 minutes with marinade. The roast will be cooked in an hour. Remove from oven, remove pork fat, and place on serving dish. When the juice in pan boils again, add 1 tablespoon flour and 1 tablespoon butter previously well mixed in bowl, hot water, and then 3 teaspoons currant jelly. Mix thoroughly, allow to boil. Add 2 tablespoons brandy. Strain sauce into sauce boat. Salt each slice of fillet as served.

DAUBE OF SLICED-BEEF NICOLETTE

A *daube* is an earthenware recipient high for its circumference. The meat that is cooked in such a recipient is always mentioned as being cooked in *daube.* Cut 3 lbs. fillet in thin slices, flatten with mallet. Cover each fillet with a thin slice of side fat of pork of the same size. Chop 3 large onions, 1/2 cup parsley, 2 cloves of garlic, 3 shallots, 1/2 cup chervil, and 4 truffles. Add 1/2 teaspoon powdered thyme, 1/2 teaspoon powdered laurel leaf, 1/2 teaspoon salt and 1/4 teaspoon pepper, 1/4 teaspoon cinnamon and 1/4 teaspoon cloves. Mix thoroughly. Spread some of this mixture on the pork fat, cover with another piece of pork fat. Recommence with the next slice of fillet-pork-fat mixture. Place these prepared slices one on top of the other and tie securely. In the *daube* or in a fireproof earthenware casserole with a cover that closes hermetically put 1/4 cup butter. Place over medium heat, add 1 cup chopped fat back of pork, chopped coarsely, the fillet, a calf's foot, a bouquet of parsley, chervil, laurel leaf, twig of thyme, 1/2 teaspoon curry powder, 1/2 teaspoon salt, 1/4 teaspoon pepper and 4 tablespoons *purée* of tomatoes. Cover with 1 bottle Marsala and 2 cups hot water. Cover and bring to a boil. Reduce heat and simmer for 4 hours. Then remove *daube* from heat. Remove meat from juice, place in bowl and remove strings when cool enough to do so. The slices should remain in place. Strain juice over slices of beef, which it should cover. Set aside in a cool place for 24 hours. Remove from bowl to serving dish.

ROSETTES OF BEEF

For six servings take 12 slices from the small end of a fillet of beef. Melt 1/2 cup butter in a fireproof earthenware casserole with a cover that closes hermetically or in an enamel-lined pot, over medium heat, and add the fillet. Heat 4 tablespoons brandy, light and pour over the fillet. Add 1 large onion, chopped fine. Cover with heavy cream and 1/2 cup strong beef *bouillon*, 1/2 teaspoon salt and 1/4 teaspoon pepper. Cover and place in preheated 350° oven for 1 1/2 hours. Place in serving dish with sauce poured over it.

VEGETABLES

ENDIVES

Wipe 12 endives, remove outer leaves, cut both ends. In a casserole with a cover that closes hermetically place 1/2 cup butter. Heat over low flame. Add the endives, sprinkle 1/2 teaspoon salt and 1/4 teaspoon sugar over them and the juice of 1 lemon. Cover with a well-buttered piece of waxed paper. Reduce heat to smallest flame, cover and cook for 5 minutes. Raise to high flame and cook for 20 minutes longer. Shake the casserole from time to time but do not uncover until ready to serve.

BROWNED SPINACH DAISY

Wash in five waters 4 lbs. spinach. Drain and press to remove most of the water. Put by handfuls in a saucepan over highest heat. Turn with a wooden spoon so that each new handful is at the bottom of the saucepan. When the spinach has been thoroughly heated cover saucepan. Gradually reduce heat to low flame. Boil for 5 minutes, drain, place under cold-water tap until spinach is tepid. Place in a saucepan over low heat and stir with a wooden spoon until all the water has evaporated. Prepare a *purée* of mushrooms by pounding through a strainer 1 lb. fresh mushrooms. Place in a saucepan over low heat. With a spoon stir until all water has evaporated. Place 1 1/2 cups thick *Béchamel* sauce in saucepan over medium heat. Add 1 cup heavy cream. Stir until thick. Add salt, pepper, a pinch of nutmeg, the mushrooms and 4 tablespoons butter. Mix well and remove from heat. In a well-buttered fireproof casserole place a layer of one-third of the spinach, then one-third of the *purée* of mushrooms—three layers of each. Cover with 1 1/2 cups *Béchamel* sauce in which 1/2 cup grated Swiss cheese has been mixed. Sprinkle over this 3 tablespoons melted butter. Stand in a recipient of hot water and put in preheated 300° oven. The water should not boil. Bake for 1 hour.

CREAMED CUCUMBERS

Peel 12 cucumbers. Dice them in 1-inch cubes. Put them in 4 cups boiling salted water over medium heat. Cover the saucepan and boil from 5 to 10 minutes, until they are no longer hard. They should not be overcooked. Remove from heat, place them under cold-water tap until they are tepid. Drain and wipe dry. In a saucepan over a low heat put 1 1/2 cups cream sauce. When hot add cucumber. Gently stir so that the sauce does not scorch. When the cucumbers and sauce come to a boil add 4 tablespoons butter cut in small pieces. Reduce heat to low flame and dip saucepan in all directions until butter is melted. Place in preheated serving dish and serve at once.

POTATOES SMOTHERED IN BUTTER

Peel 2 lbs. medium-sized potatoes, cut in eight pieces. In a saucepan over medium heat put 3/4 cup butter. When the butter has melted, put the potatoes into the saucepan and cover. Stir with a wooden spoon from time to time. Reduce heat after 1/4 hour. If the butter is too reduced add more. (This will depend on the kind of potatoes used.) Increase heat to medium, then to high. The potatoes should be browned and crisp on the outside and soft on the inside. Salt (no pepper) and serve very hot.

THE REAL RIGHT WAY FOR FRENCH FRIED POTATOES

Peel the potatoes, cut them all of the same size and length. Put them in moderately hot oil, lard or very white beef fat—there should be enough so that the potatoes are not crowded. When the potatoes come to the surface, remove them from the fat at once. Let the fat reheat quickly, increase to highest flame. The potatoes should not be out of the fat more than 2 minutes. Plunge them into the fat for the second time and remove at once. Sprinkle with salt and serve at once.

SALADS

SALAD BAGRATION

Six hearts of artichokes, in 1/3-inch slices, shredded raw celery root (celeriac). Put aside. Bring 1 quart salted water to a rolling boil. Drop into it 1/4 lb. macaroni broken into 3-inch lengths. As soon as the water recommences to boil, lower heat and cook gently for 15 minutes. Cut off heat, but allow marcaroni to stand covered for 5 minutes. Then drain thoroughly, cool. Mix the vegetables and macaroni, salt and pepper, and 1 1/2 cups mayonnaise to which

3 tablespoons *purée* of tomatoes have been added. Place in a mound on serving dish, surround with alternating borders of chopped yolks of 4 hard-boiled eggs, 8 slices chopped smoked beef tongue and chopped hard-boiled whites of eggs. Sprinkle over mound and border 2 tablespoons finely chopped parsley.

SALAD CANCALAISE

For each serving take 1 leaf of lettuce; on this place 3 tablespoons diced potato mixed with 1 teaspoon mayonnaise. On this place 3 poached oysters drained and placed on linen cloth to dry, then mixed with oil, lemon juice and pepper. On the oysters place a thin slice of truffle. Place the lettuce leaves and their garnishings on a round dish in a circle with one in the centre.

SALAD NIÇOISE

Take equal parts of diced potatoes and diced string beans. Boil separately—the potatoes in covered cold salted water brought to the boil, the string beans in boiling salted water. Do not overcook. When cooked remove from water and drain thoroughly. When cold mix the two vegetables, add two parts oil to one part lemon juice, salt and pepper. Place in a mound in salad dish and decorate with anchovies, stoned black olives and capers. Surround with quarters of peeled tomatoes. Sprinkle generously with chopped basil.

SALAD PORT ROYAL

Take equal parts of boiled sliced potatoes, shredded peeled apples and cubed boiled string beans. When the potatoes and string beans are cold, mix with apples, salt, pepper and mayonnaise. Place in salad bowl, cover with small quantity of mayonnaise, decorate with cubed string beans. Surround with quartered heads of lettuce and quartered hard-boiled eggs.

SALAD RAPHAEL

Shred lettuce and mix with mayonnaise made with 1 teaspoon paprika, instead of black pepper. Surround alternately with sliced cucumber, peeled tomatoes, cut in eight parts, and hearts of lettuce. Place between each of these vegetables 4 small finely sliced unpeeled radishes. Pour over the border a *vinaigrette* of two parts oil, one part lemon juice, salt, pepper and chopped chervil.

POTATO SALAD

In a large bowl place 1 tablespoon French mustard, 1/4 teaspoon pepper, 1 tablespoon chopped chervil, 1 tablespoon finely cut chives, 1 tablespoon finely chopped tarragon and 1 tablespoon chopped parsley. Remove the skin, head, tail and bones of 1 salt herring. Pound the flesh and mash through fine strainer, add to the contents of the bowl. Shred 1 tart apple and put in bowl. With 2 tablespoons olive oil amalgamate thoroughly, then slowly stir into it 1/2 cup strong beef *bouillon.* It should be perfectly smooth. Slowly add 1 cup dry white wine. Into a saucepan of cold water over highest flame put 2 lbs. unpeeled potatoes. Cover, and when the water comes to a boil regulate heat. Do not overcook. Remove from water, drain, peel and thinly slice while hot. Put into bowl and mix with the sauce. Keep the saucepan covered and remove one potato at a time. Be careful in mixing not to bruise the potatoes. When thoroughly mixed, put aside for at least 3 hours before serving.

SALAD MELI-MELO

Mix 10 oysters poached for 1 minute, 10 mussels, 1/2 cup shrimps, the meat of 1 medium-sized lobster, 3/4 cup very young diced string beans, 5 sliced hearts of artichoke, 10 green tips of asparagus, 1/4 cup diced beetroot, 1/2 cup celery cut in thin rings, salt and pepper. Mix lightly and pour over the mixture 1 cup olive oil and 1 1/2 cups dry champagne.

WINTER SALAD

Large sweet Spanish onions cut in thin rings with this sauce: 2 tablespoons French mustard, 1 teaspoon salt, 1/4 teaspoon pepper, 1 tablespoon paprika, 1 cup heavy cream. Mix well and pour over onions at the last moment.

SALAD LIVONIÈRE

A mixture of vegetables of the season in equal quantities. For four cups, add 1/2 teaspoon salt, 1/4 teaspoon pepper, 1 teaspoon French mustard, 3 tablespoons brandy, 3 tablespoons Madeira, 2 tablespoons oil and 1 tablespoon vinegar.

DESSERTS

Cold Desserts

BAVARIAN CREAM PERFECT LOVE

Mix 2 cups sugar and 8 yolks of eggs until lemon-coloured. Slowly add 2 cups hot milk in which 6 cloves have been heated. Put in saucepan over lowest heat. With a wooden spoon stir continuously in the same direction until the spoon remains thickly covered. Do not allow to boil. Remove from heat and pour over 1/2 tablespoon powdered gelatine that has been soaked for 5 minutes in 1/4 cup cold water. Stir in the same direction until the gelatine is completely dissolved, then strain and stir from time to time in the same direction until cool. When cold, mix with 3 cups whipped cream to which the grated zest of 2 lemons have been added. Pour into lightly oiled mould and place in refrigerator for 4 hours. Remove from mould to serving dish. The cream may be flavoured with fruit *purée.* Two and a half cups *purée* and 1 tablespoon lemon juice are mixed with 1/2 cup icing sugar.

A chocolate cream is made by melting 3 ozs. chocolate in the milk; a coffee cream, by substituting 2 cups strong coffee for the 2 cups milk. For the rest, proceed as above.

CHARLOTTE CREAM

Prepare half the quantity of the above recipe. Cut sponge fingers to fit a lightly oiled mould; for the bottom, cut to fit; for the sides, cut one end straight across, and even the tops after the lined mould has been filled with the cream. Any flavour may be used. When it is ready to be served, decorate through a forcing tube with whipped cream and surround by whipped cream. If a fruit *purée* is used, decorate on top with whipped cream and surround by whole fruits—the same kinds as used for the flavouring—cooked in a heavy syrup and well drained. Or instead of a *purée* of fruit, the recipe for Cream Perfect Love may be used, omitting the lemon and clove flavouring, and adding 1 1/2 cups fresh fruit or berries macerated for several hours in best kirsch or curaçao and well drained. A cream made with a *purée* of fruit may have added to it another fruit diced and macerated, for example a strawberry or raspberry *purée* with cubed peaches or apricots—or vice versa—the combinations are endless.*

CRÈME RENVERSÉE À LA CEVENOLE
(Chestnut Custard)

Stir 2 whole eggs and 4 yolks with 1 cup sugar until it is pale lemon-coloured. Slowly add 2 cups hot milk. Strain through a

* *Note.* Lady fingers are known as sponge fingers in English confectionery.

sieve. Stir from time to time until cold, then mix with a *purée* of chestnuts made by shelling 25 chestnuts and boiling them gently for 7 or 8 minutes. Remove from heat, drain and replace in covered saucepan. Take out a few at a time and remove skins (they peel more easily when warm). Put the peeled chestnuts in a pan over medium heat with 2 cups milk, 1 tablespoon butter, a stick of vanilla cut in half and a pinch of salt. When it commences to boil, place the covered saucepan in a 275° preheated oven. It will take about 3/4 hour. If necessary add hot milk. When quite tender, mash while very hot through fine strainer. Replace over low heat and add 1/2 cup sugar. When sugar is dissolved remove from heat. When cold, mix with egg-milk mixture. Pour into buttered mould and place in a recipient containing hot water to within 1 inch of the top of the mould. Place mould in preheated 275° oven for about 3/4 hour. It is very important that the water in which the mould is standing does not boil. If water evaporates, add boiling water. Do not remove the cream from the mould until cold.

This dessert can be poured into a ring mould, and at the last moment whipped cream lightly sweetened is placed in the centre. One tablespoon icing sugar added to 1 cup cream and flavoured with 1 tablespoon kirsch to the cup. This custard, in double the quantity before adding the chestnuts, can be prepared by adding any desired flavouring; for example, moulded coffee custard is made by replacing 1 cup milk with 1 cup strong coffee, or chocolate custard by adding 2 ozs. melted chocolate to milk.

CRÈME CARÊME

This is a dessert created by Carême, the great early-nineteenth-century cook who wrote several of the most important French cook-books.

Mix 8 yolks of eggs with 3/4 cup sugar and 1 tablespoon rice flour (or a little less than 1 tablespoon cornflour) until the mixture is pale yellow. Add very slowly 3 cups hot milk. Put in a saucepan over very low heat, stirring with a wooden spoon until the spoon is thickly coated. Remove from heat, stirring occasionally until cold. Put in the refrigerator (Carême says on ice and surrounded by ice). Remove from refrigerator from time to time to stir. Just before serving add 1/3 cup best Maraschino and 1 1/2 cups whipped cream. You will then have (these are Carême's words) a light velvety mellow cream.

APPLE MOUSSE

Peel 8 apples, remove cores, cut in eighths and put in saucepan over medium heat with 1 cup sugar, 1 cup water and the zest of a lemon. Stir from time to time, gradually lower heat. Cook until it has become quite thick. Before removing from heat add 1/2 oz. gelatine that has been dissolved in 1/4 cup water. Mix thoroughly, add the juice of 1 lemon. Strain, and with an electric beater or egg beater whip the apple sauce until it is very frothy. Put in refrigerator. Before it becomes too stiff, in about 1/2 hour, add 1 1/2 cups unsweetened whipped cream flavoured with a liqueur, Grand Marnier if convenient. Three tablespoons will be sufficient. Any liqueur is suitable, but it is preferable to use one with a fruit base rather than a herb. Serve chilled in bowl.

FRUIT MOUSSE

One lb. mixed fruits of the season—bananas and fresh pineapples are always acceptable additions. Peel the fruits and remove stones, cut bananas in half their length and half their width, pineapple in thin slices—do not dice—and leave large fruits whole. Arrange in deep serving dish and place in refrigerator several hours before serving. At the last moment mix 8 yolks of eggs with 3/4 cup best kirsch and 1 cup sugar. Beat with an electric beater or egg beater until thickened and frothy. Pour over fruits and serve.

APPLES WITH RASPBERRY SYRUP

Peel, core and cut in quarters 1 1/2 lbs. apples. Put in saucepan over high heat with 1 cup sugar, 1/2 cup water and the zest of a lemon. Stir to prevent scorching. Boil for 20 to 25 minutes. Remove from heat and strain into lightly buttered mould, the bottom of which should have a piece of white paper to fit. When cool place in a refrigerator for several hours. Unmould when ready to serve and turn on to serving dish. Pour over it 2 cups strawberry or raspberry *purée.* If these fruits are not in season, heat 1 1/2 glasses jelly with 2 tablespoons water. Heat only sufficiently to liquefy the jelly, which should be chilled before using.

PEACHES GLACÉES

Put 6 fine peaches in hot water only long enough to peel. Prepare a syrup of 1 cup sugar and 3/4 cup water. Poach the peaches covered in the syrup over low flame for 4 minutes. Remove peaches and drain. When the syrup is quite thick, after about 3 minutes' further cooking pour over peaches. See that it adheres. When cold place peaches on serving dish in refrigerator for at least 2 hours. Prepare a *purée* of 1 lb. fresh strawberries, 1/4 cup icing sugar, add 2 tablespoons best brandy and 1 cup whipped cream. Put in refrigerator for at least 2 hours. Before serving pour *purée* over peaches.

Hot Desserts

APPLE FRITTERS

Remove core and peel 6 apples, cut in 1/3-inch slices. Put in bowl and macerate for 1/2 hour in juice of 1 lemon, 2 tablespoons sugar and 3 tablespoons brandy. Turn several times in order to moisten and sweeten all the slices. The batter is prepared by sifting 1 cup flour into a bowl and mixing gradually with 1 egg, 1 cup light beer, 3 tablespoons brandy and 3 tablespoons olive oil. Mix thoroughly, the batter should be perfectly smooth. It is best, if convenient, to prepare the batter the night before using. It must be made at least several hours before being used. The apples are dipped in batter and at once fried in deep fat. When browned, remove with perforated spoon, cover generously with sugar. While still hot, place in hot oven to glaze.

All fruits are prepared in this way for fritters.

OMELETTE À LA BOURBONNAISE

Prepare an omelette in the prescribed way with 6 eggs and 1 tablespoon sugar. Before folding, place on it 1 cup diced pineapple mixed with 1/4 cup strained apricot jam. Fold and sieve 1/4 cup sugar over the omelette. Pour 1 cup rum over it and light in the dining-room before serving.

This type of omelette can be filled with 1 cup strained jam and any liqueur suitable to the flavour of the jam; for example, strawberry with kirsch, or orange with curaçao.

FRUIT SOUFFLÉ

Prepare a heavy syrup with 1 cup sugar and 1/4 cup water. Let it boil for 5 minutes. Mix with 1 1/2 cups *purée* of any fresh fruit. If it is not thick boil to reduce and then measure. This would be necessary with berries. Mix lightly but quickly with 6 whites of eggs that are beaten at the last moment—do not let them stand. Turn into a buttered and sugared *soufflé* dish and bake in preheated 375° oven for about 20 minutes. After 18 minutes sieve icing sugar over the top to glaze.

VIOLET SOUFFLÉ

Melt 2 tablespoons butter in a saucepan over medium heat, add 2 tablespoons flour stirring constantly. Add 1 cup hot milk and boil until thick. Remove from heat and cool to tepid. Add 2 tablespoons sugar and 1 tablespoon Kummel, and the yolks of 3 eggs, one at a time, mixing each one thoroughly before adding the next. Beat 4 whites of eggs and lightly and quickly add to mixture. Add 1/2 cup candied violets broken into three or four pieces each. Place a wreath of candied violets on the top of the *soufflé*. Pour into a *soufflé* dish and place in preheated 375° oven for 20 minutes.

CROÛTE À LA BELLE AURORE

Cut stale cake in slices 1/3-inch thick. Sprinkle them generously with rum. Cover half of them with very thick apple sauce to which has been added 1 tablespoon heavy cream to each cup of apple sauce and 3 tablespoons tiny dried mixed *glacé* fruits. Place the other slices of cake on the prepared ones, pressing so that they hold together. Dip in beaten egg, cover with rolled sifted cracker crumbs and fry in butter over medium heat. Serve with

Hot Sabayon Sauce

For six servings, stir in saucepan over lowest heat as in bain-marie 3 yolks of eggs with 8 tablespoons sugar until thoroughly mixed. Add very slowly 2 cups Marsala, beat with a whisk until it thickens. Serve in preheated bowl or sauce boat.

X.

SERVANTS IN FRANCE

Unfortunately there have been too many of them in my service. Unfortunately there have been too many unsatisfactory ones, and too many of the satisfactory ones did not stay long. My housekeeping experiences in Paris date from 1908 and during forty-five years many servants come and go for many different reasons. The memory of some of them is vague, even of some of them who were satisfactory and stayed for quite a long time—that is, comparatively a long time. But my memory is quite vivid of some who came and left within a day or two and that perhaps is natural.

The first servant I engaged left a pleasant memory. A friend and I took a flat, small, sunny and cheerful, in 1908. When the furniture was being moved into it the time had come to think of a servant who would take care of it. When I asked Gertrude Stein how you went about finding one in Paris and what questions you put to her in your interview with her, she answered that it was easiest and most reliable to find her through connections in the quarter, and the proper question to ask was, did she make a good omelette. The next day a nice young girl, a Swiss-Italian, presented herself. She was the niece of Gertrude Stein's *concierge.* She would therefore be honest and respectable, as one had reason in those days to suppose, and she said she made a good omelette. So she was engaged. Within a few days we moved and Célestine, sweet, smiling and timid, was there to greet us and to begin her work. When it was time for her to go marketing and she was given a list of what she was to buy, together with some money and an account book, she confessed that she had never marketed. She was more discouraged than I. We bought the provisions together. If she cooked well enough experience would do the rest. Lunch was inedible—commencing with the omelette. She obviously had not the slightest instinct for cooking. I told Gertrude Stein why we could not keep her and together we told Gertrude Stein's *concierge*, who said that perhaps she should take a position as a lady's maid.

Then a friend recommended a middle-aged Basque cook who would like to cook and do the household work for two ladies. Maria Lasgourges was a treasure, an excellent cook, resourceful and experienced. All that winter we had guests, invited, unexpected and uninvited, who were served with Maria's unfailingly delicious food. She had an elaborate, formal, ceremonious manner towards the bohemian acquaintances we had met. She only made one blunder: we were having six guests for Christmas dinner and Maria was told to cook a dinner suitable to the occasion and the position of our guests. She was left as usual to select her menu without consultation, but was warned that she should buy anything that she needed to cook a good dinner. What was our surprise and mortification to find that the first dish was cold boiled crawfish in their shells—exactly five for each person. What followed was chosen and cooked with her usual taste and competence. When I thanked Maria and complimented her after the guests left, the crawfish were not mentioned. They were spoken of the next morning as a surprise to us. She said she had expected they would be, that she had considered the expense but concluded that they were indispensable for a Christmas dinner.

When in spring we had planned our summer vacation we spoke to Maria and told her that we proposed paying her her wages during our absence. Perhaps she would go to stay with her people in the country. She thanked us and said she was getting too old to work and had planned to go to them permanently. It was a blow to us. She had been an expert cook and housekeeper, and a warm and kindly presence in the flat.

When we returned to Paris an acquaintance leaving Paris asked us if we didn't want her perfect servant for whom she was hunting a suitable situation. Maria Entz was Swiss, with all the Swiss virtues and limitations. She was clean, hard-working and honest, and her cooking was indifferent, either too heavy or too economical. We suffered from her incurable romanticism. She liked to cook dishes in which cheese or chocolate were the principal ingredients, and wrote to Switzerland to have them sent to her. She made honey cakes cut out to represent vegetables, fruits and flowers, men, women and children, birds and animals, garden implements and kitchen utensils. With these and her unvarying reliability we had to content ourselves and not too greatly regret the marvellous other Maria. Once more it was time to announce our plans and conditions. Maria Entz told us that in any case she was about to be married (blushes and giggles), having completed the purchase of her personal and household linen. We were surprised, as we had considered her a confirmed old maid. We told her that we hoped that she would be happy and that she was marrying a man who would be worthy of her. With more blushes and giggles she said he would prove so as he was young and she could form him—at twenty-two that could easily be managed. For her wedding gift a veil with a wreath of roses was suitable.

After that vacation of 1910, I went to live with Gertrude Stein, whose admirable servant Hélène had charge of her and her household. It was Hélène who made all the practical decisions. A friend noticing this observed that it was to be hoped that her servant left a free choice of the Picassos to Gertrude Stein. No one could have accomplished so much with so little effort and in so little time as Hélène did. She was that rare thing, an invariably perfect cook. She knew all the niceties of making menus. If you wished to honour a guest you offered him an omelette *soufflé* with an elaborate sauce, if you were indifferent to this an omelette with mushrooms or *fines herbes*, but if you wished to be insulting you made fried eggs. With the meat course, a fillet of beef with Madeira sauce came first, then a leg or a saddle of mutton, and last a chicken. Elaborate desserts she said were not made by plain cooks, and to her any dessert more complicated than a *soufflé* was that. When she saw the cake I ordered for tea from an amazing Viennese baker, she asked me to order a dessert from him for a dinner party. She was horrified but impressed when she paid for it, and as proud as if she had made it when it was served.

Hélène was without humour and was always practical. When the *Titanic* sank, she said she thought the Anglo-Saxon gallantry of saving women and children first was unintelligent and unnatural. She believed families should be saved first, then single people. What, she asked, are widows to do without husbands to take care of them. She supposed that there would be no painters in the United States, since they all came to Paris to learn to paint. When I asked her what she thought Americans did do, her answer was that most of them must be dentists. From Hélène I learned nothing about cooking. She would have thought such an idea was misplaced. A lady did not cook. In the spring of 1914, Hélène's husband said she should no longer work for others, but stay at home and take care of him. It would be a sad parting for Gertrude Stein, who had had Hélène's admirable care and consideration for nine years.

Eventually a good cook was found—Gertrude Stein at once nicknamed her Muggie Moll—a handsome woman with a cast in one eye which she cleverly used to her advantage. Her husband was a *gendarme*, which was a guarantee for her honesty and respectability. We left for a visit to England just before portentous Sarajevo occurred, and then were unable to return to France until after the Battle of the Marne. Muggie Moll, whose husband was away at the war, in spite of the difficulties, managed to find suitable food. Then she suddenly announced that for reasons over which she had no control she would have to leave.

A *concierge* in the quarter recommended a young woman whom she had known for several years. Jeanne was a suitable person for us, she had worked for six years in a family and had lately been dismissed, the reason for which would not be repeated. She advised us to take her. Jeanne was sent for and proved to be a treasure. She was a very good cook, kept the *pavillon* (our garden house) scrupulously clean and was attentive to our wants. Our guests those days were frequently unexpected. Jeanne was adequate in emergencies.

One day when she sighed, I asked if she was worried. She explained that, Belgium being occupied by the wicked Germans, we would have no endives that winter, since the best endives were imported from Belgium. When a Zeppelin raid had sounded, Jeanne, pointing to the sky, said that something soon would be pelting down on us. To be casual I said, perhaps not on us. But, Jeanne said, such heavy black clouds meant we were in for a rainstorm.

When we commenced to work for the French wounded, we left Paris to visit hospitals and Jeanne cleaned the *pavillon* from top to bottom. When we got home after one of the trips she told us she was to be married. The war had changed one's attitude to many things. Losing Jeanne was not the tragedy it might have been. But during many years I regretted her.

Then our history with *femmes de ménages* commenced. It was a long and not always a happy one. Léonie, the first of them, came to us very soon after peace was signed. Too hardworking for her years, her boundless energy and reckless courage frightened me. With her long arms and frail body she would climb a stepladder with the agility and the unnatural smile of a professional acrobat. In her brusque way she asked me one day if she might be allowed to make us some cakes. More curious than hopeful I agreed. She said they were named after the nuns who first made them long ago. The religious order to which the nuns belonged was that of the Visitation, and the cakes, like the nuns, were called

VISITANDINES

Put 1 1/4 cups butter in a saucepan and allow it to become golden brown. Put aside and cool. Add one by one to each other the unbeaten whites of 6 eggs, adding them very slowly and stirring with a wooden spoon. This should take 1/4 hour. Then add 2/3 cup sifted flour. Fold in until perfectly smooth. Add 1 teaspoon vanilla extract and the cooled melted butter. Fold in 2 beaten whites of eggs. Fill small buttered muffin pans and bake in preheated

400° oven until pale gold. When baked, transfer to grill and paint with reduced strained apricot jam. Cover with kirsch icing.

They were delicate and tasty and made evident Léonie's possibilities. If she could bake as well as the *Visitandines* proved she could, was she perhaps an equally good cook. As she was engaged to do the heavy work, it was not for me to ask her if she could cook until she suggested it, which she did. It did not take long to appreciate its quality. We had certainly luck in finding good cooks, though they had their weaknesses in other ways. Gertrude Stein liked to remind me that if they did not have such faults, they would not be working for us. The first dish Léonie prepared for us was

BAKED ZUCCHINI OR ITALIAN SQUASH

Wash and dry but do not peel the squash, preferably small and definitely firm. Cut them in half lengthwise. With a sharp knife, cut the flesh all around within 1/4 inch of the skin but not at the bottom. Cook in boiling water, salted, from 4 to 6 minutes according to their size. Drain well and wipe dry. In a frying pan melt 1/4 cup olive oil and cook in it 4 medium-sized onions, for 4 medium-sized squash. Remove the flesh from parboiled squash without piercing the bottom, chop fine. Add to onions with 1/2 cup fresh breadcrumbs and 2 cloves of crushed garlic, salt and pepper. Mix well, remove from flame and add 2 tablespoons chopped basil and 1 tablespoon chopped parsley. In 10 minutes add 2 well-beaten eggs. Mix well and fill squash shells with mixture. Place in well oiled, fireproof dish. Cover the squash with a thick coating of dried breadcrumbs. Sprinkle generously with olive oil. Bake for 1/2 an hour in preheated 350° oven. Baste two or three times during baking.

After we enjoyed the squash Léonie gradually came to do the greater part of the cooking and I did more of the household work.

Léonie was physically unable to work longer than she originally proposed, nor could we have afforded it. She continued to be a good reliable cook, and asked me to give her some recipes, which I did, but it was obvious that she did not measure or weigh the ingredients. She preferred to pray. To tease her I told her she did not always pray to the right saint.

After two years had passed, we felt that fortunately Léonie had become a permanency. But before the end of the third year she had gone, and it must be acknowledged that later we could not remember the reason why, though we were far from ungrateful to her for her faithful service.

Then a pearl graciously condescended to work for us. This second Jeanne was indeed without price. She was for days uncertain what she should be paid, and we were equally certain no amount would be too much. Jeanne was gentle and had the smile of a *cocotte*, wore a complicated headdress and was as naïve and inexperienced as a baby. At the end of a week, we came to an arrangement for the wages she was to be paid. It was then that she said that she would, if it suited me, tell me each morning what the previous day's accounts had been. To "tell them" apparently meant to "give them." In a flash her hesitancy and some of her mystery were explained. She neither read nor wrote. After that she told me what she had spent the previous day by placing the money that remained on a table and asking me if she had done the account correctly. She was attaching. In her very lovely low voice she said little. She was strange, but agreeably so. Gertrude Stein was unable to interpret Jeanne's mystery.

We were becoming acquainted simultaneously with Jeanne and her cooking. *La cuisine c'est la femme.* Her sauces had unknown, delicate and still exotic flavours. A dish would have an unexpected squeeze of orange juice. She used rum in all her desserts. When she prepared chicken with fresh coconut, we had an answer to some of the strangeness. Pondicherry. And we were satisfied. And she pleased us more than ever, and well she might, when she made us such dishes as

CREAM TART

Stir the yolks of 3 eggs, 1/4 cup sugar and 1 tablespoon flour until they are pale yellow. Add slowly 1 cup hot milk. Put in saucepan over very low flame and stir constantly until thick. It must not boil. Strain through fine sieve. Line a 9-inch pie plate with puff paste. When the egg-sugar mixture is cold, pour it on the crust and cover with

Coco Marmalade

Over low flame melt in a saucepan 2 cups sugar and 1 cup water. Stir until it commences to boil. When it begins to thicken, add 1 1/2 cups freshly grated coconut or moist grated coconut. Boil until transparent. Pour over cream in pie crust and bake in preheated 350° oven until cream is stiff and the top has browned.

It is still one of our favourite tarts. But was it one from Pondicherry?

We were commencing to feel that that was not the answer to anything relating to Jeanne's strangeness. We didn't know until she had been with us for seven or eight months. Then one morning she did not come. A *pneumatique* explained her absence. She had been suddenly ill during the night, she had a high fever and as soon as the doctor had seen her there would be another bulletin. The following day there were two more which told nothing. We strolled down on the third afternoon to see her *concierge* who said briefly that Jeanne was seriously ill and would not be able to see us, but we would hear how her illness was progressing. More bulletins rained upon us. At the end of the week, the *concierge*, strangely embarrassed, said Jeanne's condition remained unchanged. Just as we were about to question the *concierge* further, in bounced Jeanne as gay as a lark, her arms filled with packages of all sizes and shapes and a bandbox balanced on one finger. Of the four of us the only one who had any presence of mind was the *concierge.* She wafted Jeanne out as she told her to go to her room and she would explain all. It was a bare and plausible story.

Twenty-odd years ago Jeanne had rented the room and kitchen she was still living in. She did not work then, she received letters from Martinique—not Pondicherry—for some years. Then they ceased. The people who came to see her were well behaved, there was neither noise nor disorder from Jeanne's room. Then she commenced to work. After several months' work she would stay in her room and not take on new work for a month or even longer. Always during this time, she would go out at night and alone.

We told the *concierge* that we liked Madame Jeanne, as we had been speaking of her, and would like her to return as soon as possible. The same evening there was word saying that Jeanne would return to us the following Monday morning. When she did, nothing was said on either side of her illness or the return from her shopping expedition. We could once more enjoy her lovely low voice, her gentle ways and her *cocotte* smile. She prepared her subtly flavoured food which was neither from Pondicherry nor Martinique, but her very own. And to please us and to vary the menu she cooked us

CHICKEN CROQUETTES AND EGGS

Put through the meat chopper enough chicken, skin removed, to make 4 1/2 cups. Season with salt and pepper, a pinch of cayenne, and 1 crushed clove of garlic, 1/4 teaspoon powdered mace, and 2 tablespoons cream cheese. Form into round flat *croquettes* with slightly floured hands. Fry in 4 tablespoons butter, brown on both sides, lower heat the last 5 minutes, and pour over the *croquettes* 6 well-beaten eggs. As much as possible, prevent the eggs from touching the pan. They should cover the *croquettes* and only cook enough not to run, for about 3 minutes. Turn the pan upside-down on a preheated dish. Serve hot with

Sauce for Croquettes

Put 1 1/2 cups chicken stock in a saucepan over low flame with salt and pepper, 1/4 teaspoon mace, a pinch of cayenne pepper, add 2 tablespoons *purée* of tomatoes, mix well. Then add 3 tablespoons cream cheese, heat well but do not allow to boil. Strain and pour into sauce boat and serve with *croquettes.*

Jeanne one morning did not appear to make my coffee, but this time when she did not come to her work there were no bulletins concerning the state of her health. On the second afternoon we went to her *concierge* and told her to say to Jeanne that if she did not return the next morning she should not bother further about us. The following morning, when she had not appeared, I went sadly to find someone to succeed her. It was no longer so easy to find a satisfactory servant. It was not a surprise not to have been able to engage one, but it was a surprise to find Jeanne bringing my coffee the next morning. There was no reference on either side to her A.W.O.L. Jeanne had her lovely low voice, her *cocotte* smile and her gentle ways in spite of her shopping expedition having been curtailed.

Jeanne's cooking was always varied. She smilingly, not boastingly, said she supposed she could cook eggs and potatoes in a hundred different ways. Here is an example of what she did with eggs:

MIRRORED EGGS À LA BRAHAN

For 4 eggs, chop 1/4 lb. mushrooms, cook with 2 tablespoons butter, salt and a pinch of lemon juice in a covered saucepan over low flame for 8 minutes. Chop or put through meat grinder enough chicken to make 1 1/2 cups. To this when ground add half its weight of butter, about 5 tablespoons, and salt and paprika. Make miniature small round *croquettes* and fry lightly on both sides in 2 tablespoons butter. In well-buttered fireproof dish spread the mushrooms; place 4 eggs on the mushrooms and 1 teaspoon butter on the yolk of each egg. Place the dish in preheated 400° oven on asbestos mat for 5 minutes. The chicken *croquettes* must be ready and hot, as well as 1 cup cream sauce to which the yolk of 1 egg has been added but not allowed to boil. When the eggs are cooked, remove the dish from the oven, place the chicken *croquettes* against and around the eggs, and pour a narrow border of cream sauce around them.

This is another and equally original way Jeanne prepared eggs:

POACHED EGGS À LA SULTANE

Bake puff paste in fluted *pâté* shells. When baked and still hot place in each one a poached egg. Cover with a sauce made in this way:

For 6 *pâté* shells, melt 1 1/2 tablespoons butter in a saucepan over low heat. When butter is melted add 1 1/4 tablespoons flour. Turn with a wooden spoon until thoroughly amalgamated, then add slowly 3/4 cup strong hot chicken *bouillon*. Stir constantly over lowest heat for 5 minutes. Add 1/2 cup heavy cream. Do not allow to boil. Add 1/4 cup pistachio nuts that have had their skins removed by soaking for 3 minutes in hot water. Dry and rub in cloth—the skins will loosen and finally remain in the cloth. Pound them in a mortar with a drop of water added from time to time to prevent the nuts from exuding oil. When they can be strained through a sieve, add 1/4 cup and 1 tablespoon soft butter to them and mix together. Add this mixture very slowly (called, naturally, pistachio butter) to the chicken *bouillon* cream sauce. Heat thoroughly but do not boil. Cover the eggs with this and serve at once. As good as it looks.

Here is one of the many ways Jeanne cooked potatoes:

POTATOES CRAINQUEBILLE

Chop 2 large onions. Melt 4 tablespoons butter in a saucepan over low flame and cook the chopped onions in it without letting them brown. Put them in a shallow fireproof earthenware dish. Cover them with large new potatoes. Pour 1 teaspoon melted butter on each potato. Cover the dish and cook on an asbestos mat over low flame for 1/4 hour. Then put in preheated 350° oven for about 1/2 hour depending upon the size of the potatoes. When they are tender enough to indent with a fork remove from oven. Turn the oven to 450°. Gently press on each potato to make a small hollow in the centre. In this, place 1 teaspoon of tomato *purée* to which a quarter of its volume in butter has been added. Cover the *purée* and potatoes with grated Parmesan cheese and return to the hot oven to brown. This too is a dish as good as it looks.

One more of Jeanne's recipes, for

POTATOES MOUSSELINES

Bake 2 lbs. potatoes in the oven. When cooked for about 3/4 hour in a hot oven, according to size, remove skins, mash through strainer, add 1 teaspoon salt, 1/4 teaspoon pepper, a pinch of nutmeg, 3/4 cup soft butter and 4 yolks of eggs. Mix thoroughly. Add 1/2 cup whipped cream. Cover with melted butter and put for 5 minutes in a 450° preheated oven.

Jeanne continued to be a delightful presence in the *pavillon* and a pearl of a cook. Then the inevitable occurred. One morning and for the third time she did not appear. When Gertrude Stein saw me bringing her her breakfast tray, she knew what had happened. We said in chorus, Never once without twice and thrice. We sent word to Jeanne's *concierge* that we were making other arrangements and did not expect Jeanne to return. For a long time afterwards I went out of my way to pass where she lived on the chance of seeing her, but never did. She was buried in one of the department stores making her purchases. I always hoped she had gone across town for something more satisfactory than window shopping.

After Jeanne we had no success in finding a suitable servant. The French law permits one to engage a servant and dismiss her the next day on the condition that she is paid a week's wages and living expenses, or she may be told that she will be dismissed in a week and her wages paid for the week. Experience taught me the arithmetic for these calculations, and it was at the pencil's tip. One of the servants who were coming and leaving continuously surprised us. On her arrival early in the morning she said that with my permission she would give a thorough cleaning to the kitchen before she undertook any other work, and would I please tell her where the stepladder was. Later in the morning when I saw how hard she was working, I suggested to Gertrude Stein that we go to a restaurant for lunch. We could safely leave her alone. She had been highly recommended by our butcher. She was given some money to market and for the household material she might need. She impressed us favourably. She opened the door for us and then I remembered to tell her she should not attempt to clean the studio, not to derange anything in it. We came home later than we had expected to. The door was opened by a person whose expression was severer and more resolute than I had remembered from the morning. Firmly but not aggressively she announced that she was leaving at once. She found the conditions regarding her work were different from what she had supposed they would be. We did not understand. A friend said the decision was come to after seeing the pictures in the studio. They had frightened her.

The reasons they did not wish to stay, given by some of those that followed, were strange. One young woman seemed to be both satisfactory and satisfied. She was a *gourmet*, which should include being a good cook. She was not, but enjoyed the cooking I did. This did not flatter me, it bored me. She asked me if she might eat her meals before she served us. It was agreed that she should do so. The first time we had guests was for Thanksgiving Day, a lunch for two out-of-town American boys and their father. The traditional menu was to be cooked. The turkey was an imposing bird. About three-quarters of an hour before lunch was to be served, when I was basting it, Louise came to the oven with a carving knife and fork. She said that a wing or a second joint would be sufficiently cooked for her to eat her portion at once. She had commenced her lunch and was now ready for turkey. It was necessary to make clear to her that the turkey would not be presented and carved at table with even a small amputation. She threw her knife and fork on the table and burst into tears, sobbing that it was a cruel thing to do to her. It was not a precedent but an exception that Louise failed to understand. So she departed.

When an Austrian agreed to come to us, recommended by Americans from Dayton, Ohio, for whom she had worked first in Paris and then in Dayton, I told her we liked American cooking but did not wish grilled steak and chops more than once a week each. She was a nice person and apparently pleased with her work and us. It had come to our being pleased when they were pleased with us. On the third morning she looked at me severely and said that we "lived French" and that that was not what she had been led to suspect and she was leaving, which she did.

Then three very nice Bretonnes, sisters, came in turn to be our servants. Their cheerful, happy point of view toward their work encouraged me to hope that we were returning to the quiet and contented household of other days. They all three were attractive, gay, intelligent and responsible. Jeanne, who was the first to come to us, was the prettiest, the eldest and the best cook. The three of them each had gone to a convent school, but Jeanne had received the best and longest education. She spoke an almost classical French, and she cooked the classical French kitchen, so we were not surprised when she prepared

BROWN BRAISED RIBS OF BEEF

Ask the butcher to remove the bones from 3 ribs of beef and chop each rib bone into two or three pieces. Place the meat in a Dutch oven and brown in 4 tablespoons butter on all sides. Remove the meat when well seared and brown the bones. Chop 1 hard-boiled egg, 1 cup raw mushrooms, 2 large stalks of celery and 1 stalk of parsley, and 3 tablespoons fresh pork fat. Place these in the Dutch oven with salt and pepper. Place the meat on aluminium foil. Press the mixed chopped ingredients on top of the meat. Wrap securely in the aluminium foil. In the Dutch oven place a 1/2-inch slice of back fat from which the skin has been removed. Put the wrapped meat and the bones on this, add 1 cup of hot *bouillon* and 1 cup hot dry red wine, 1 medium-sized onion, salt and pepper, a bouquet of parsley, thyme and laurel. Simmer covered over low heat. After 1 1/4 hours, turn the meat and simmer 1 1/4 hours further. It may be necessary to add a little more hot *bouillon* and wine, half and half. After 3 hours' cooking, remove from flame. Place meat on preheated earthenware dish. Strain the juice and skim thoroughly. Pour over the unwrapped meat. Press 1 cup fresh breadcrumbs, 1/3 cup grated Swiss and 1/3 cup grated Parmesan cheese and 4 tablespoons melted butter, well mixed together, on the top of the meat and put in the oven at 450° to brown.

This is a luscious and savoury dish, but be warned that it has nothing to do with roast beef.

With Jeanne in charge of the household, with our usual resilience we soon forgot the dark days we had suffered before she came, and with a baseless optimism accepted our present satisfaction as permanent—until Jeanne told us that she was pregnant and would not be able to work more than a couple of months more. We were not disturbed by this. She had written to her sister of whom she had spoken to us, and to whom she had said that she should be prepared to come

and replace her. Caroline had a heart of gold and was in every way superior to her, Jeanne. We would see. There was nothing to do but accept. In the evening Jeanne's husband would call for her. He made a feeble joke by saying that if he hadn't married Jeanne he would have married Caroline, and Jeanne added that among six unmarried sisters he had ample choice. Finally Caroline arrived, and Jeanne stayed in her little home to wait for her baby to come.

Caroline was not pretty, attractive and charming, but she certainly had a heart of gold. We always spoke of her to each other as Heart of Gold. She was perpetually sacrificing herself at the expense of her state of mind or her pocketbook. Their prettiest sister, it had been discovered, was living with a married man. The family was outraged. Such a thing had never happened amongst them before. Caroline persuaded one sister and brother after the other to forgive her. For a much younger sister, she had supplemented the wages of an inexperienced worker from those she herself received, and when the sister's wages were raised continued to give her the supplement. Such was Caroline who came to work for us.

She quickly learned our ways and even Jeanne's, and her own were intelligent and amusing. She told many stories of the life of the Breton village where she had been born and raised, of what could and did happen to the boys and girls and the married couples and the aged. She said the solutions of these stories was a question of souls or *sous*.

As they had been deprived of sugar during the war, she made us at once two sweets. This is a very simple liqueur to prepare.

NOYEAU

Mash with a heavy hammer the stones and kernels of 16 peaches. Put in a jar and cover with brandy. Allow it to infuse for 1 month. Filter through a fine linen cloth. Melt 1 lb. sugar over lowest heat with 1/3 cup water. Stir until it is about to boil, then tip the casserole in all directions. Allow to boil for 2 minutes. Remove from flame and skim. Add the filtered brandy, and bottle.

This is a delicious liqueur. To make an equally good curaçao use the same recipe replacing the peach stones and kernels with dried mandarine orange or tangerine peels and allow to infuse for 2 months.

KALOUGA

Place 1 cup cream and 1 cup sugar in enamelled saucepan over low flame, stirring constantly until the mixture is the colour of coffee and cream. Then remove from flame, pour on oiled marble or enamelled sheet. Cut, before they cool, into squares or diamonds. This is a simplified and rich version of fudge, and proves the commonplace that there is nothing new under the sun.

And this was the best strawberry jam we ever had on our shelves or table and the simplest and quickest to make.

STRAWBERRY JAM (1)

Take equal weight of carefully chosen berries and sugar. Macerate for 12 hours. Then bring to a boil. Boil up three times and then quickly remove berries with flat perforated spoon. Replace syrup over medium heat, skim completely, boil until a spoon is lightly coated. Remove from flame and cool. Put berries in glasses and cover with warm syrup. Cover and store. The berries prepared in this way retain all their flavour.

Caroline could cook other dishes than sweet ones and did during a long hot summer that we devoted to getting the *Légion d'Honneur* for an American friend who highly deserved and ardently desired it. Her work for France had not been spectacular, it was known in fact to only a few friends. We needed someone with influence. We sometimes thought an American colonel could and would do it for us, sometimes a French judge, or a Hungarian man about town, or just anybody whom we had met even for the first time. We gave a lunch party for each one who looked likely. By July it was very hot. We kept the dining-room cool. It was blisteringly hot in August. The dining-room was cool, but the kitchen was insupportable. Caroline was stoic. We bought her a deck-chair with cushions and installed it in a corner of the dining-room where she could rest when she was free.

To prove that she could prepare more than the simplest dishes of meat for these lunch parties, she cooked

A DUCKLING WITH PORT WINE

Put 24 fresh figs covered with the best port into a hermetically closed jar. Let macerate for 26 hours. Then put the duckling into a preheated 400° oven in a well-buttered fireproof earthenware dish. After 15 minutes put the strained figs around the duckling and commence to baste with the hot port wine in which the figs macerated. Baste every 15 minutes. If there is not enough port wine baste with veal *bouillon* or chicken *bouillon.* To roast the duckling, 3/4 hour is ample time if it is medium sized.

To be served with rice, potatoes and green peas.

She also made a very good and attractive vegetable dish.

PURÉE OF CELERY ROOT WITH CREAM SAUCE

Very early celery root (celeriac), when it is still small and tender, is peeled, mashed, cut in thick slices and boiled in salted water for 15 or 20 minutes according to their size. Those Caroline used were so small she allowed 2 for each person. As soon as they are tender, remove from heat. Drain thoroughly and strain. Pound with potato masher through a sieve. Replace in casserole over low flame and add 1/2 cup heavy cream to 4 small celery roots. Cook, stirring constantly until reduced to consistency of a thick *purée*, for about 1/4 hour. Place, in the form of a dome, on a preheated serving dish, and cover with 1 cup heavy cream sauce for 4 celery roots. Sprinkle thickly on this skinned pistachio nuts lightly browned in the oven and then cut in fine slices. They are of course prepared in advance. It does take a long time, but they add a flavour and are very pretty.

Caroline continued faithfully to produce these lunches. In August, when Gertrude Stein and I were forced to acknowledge our efforts for our friend had failed, Caroline went to Brittany and we went to Normandy to cool off. Caroline rested, but not for long. The youngest sister, who received the supplementary wages, had come to Paris to work, and Caroline chaperoned her several evenings a week to public dances for Bretons where Margot danced tirelessly. This exhausted her considerably older sister. Caroline however remained our conscientious servant.

That winter, uneventful in the kitchen and household, Margot would call for her sister and they would go off to see the sights of Paris. Then Caroline heard from the sister who was the friend of the married man that perhaps as he was now a widower they might be marrying. Cécile came to Paris to see Caroline. What was expected of Caroline was evident. She would take care of the baby. So Margot, the third of the sisters we were to know, was sent for to work for us.

Margot was a great favourite with everyone who came to see us. She cooked well and quietly. It was she who could toss *crêpes* high in the air and light a flaming dessert as she carried it from the kitchen to the dining-room. She would have arranged the flowers if she had been encouraged, and with taste, no doubt, for the dishes she served were effectively and appetisingly presented. She often made an

ALEXANDRA SALAD

Boil 1 1/2 lbs. potatoes in their skins until they are just tender—do not overcook. Boil in salted water 1 1/2 lbs. carrots.

Shred the white stalks of 1 head of celery. When the potatoes are tender, peel and cut in thin slices, and the carrots likewise. Put aside 1 potato and 1 carrot. Mix the three vegetables in a bowl. In a small bowl, mix 1 1/2 tablespoons vinegar, 4 tablespoons olive oil, 1 teaspoon salt, 1/2 teaspoon pepper, 1/2 teaspoon dry mustard. Pour this over the sliced vegetables and macerate for 1 hour. In another bowl place 1/2 cup diced lean ham, 1 cup mushrooms that have previously been boiled for 6 minutes in water to cover with 1/4 teaspoon salt, a squeeze of lemon and 2 grated apples. Mix the contents of the two bowls and 1 cup very stiff mayonnaise. Place in salad bowl in a mound. Sprinkle liberally with chopped parsley, tarragon and finely cut chives. Around the base decorate with alternate slices of the carrot and potato that were put aside.

Margot had not much actual experience of cooking and was pleased to learn. The three sisters all had an instinct for cooking. Margot in time would equal Jeanne. Margot went with a cousin to the Breton balls so successfully that finally, with the approval of Caroline who had come down to Paris expressly, she married a young man she had met there. For the three years the sisters had served us we were carefree and very content. Gertrude Stein used to say nothing seemed more unnatural to her than the way a servant, a complete stranger, entered your home one day and very soon after into your life and then left you and went out of your life. This was not literally true of the three sisters.

When it was evident that connections in the quarter were no longer able to find a servant for us, it was necessary to go to the employment office. That was indeed a humiliating experience, from which I withdrew not certain whether it was more so for me or for the applicants. It was then that we commenced our insecure, unstable, unreliable but thoroughly enjoyable experiences with the Indo-Chinese.

Trac came to us through an advertisement that I had in desperation put in a newspaper. It began captivatingly for those days: "Two American ladies wish—" There were many candidates. Trac was my immediate choice. He was a person with neat little movements and a frank smile. He spoke French with a vocabulary of a couple of dozen words. He would say, not a cherry, when he spoke of a strawberry. A lobster was a small crawfish, and a pineapple was a pear not a pear. The Chinese cooking was delicate, varied and nourishing. To see Trac, immaculate in white, slicing in lightning quick strokes vegetables and fruits was an appetiser. There was only one course in which he was weak. He made very few desserts and those were of the simplest. To his childish joy, I taught him several. But before this he served one evening a very dubious elaborately frosted and decorated cake. There was something familiar about the cake. Did you make the cake, Trac, I asked him. As he answered that he had, I remembered that I had seen it for years, or one like it, in the window of a very second-rate confectioner's. Are you sure, I ruthlessly continued. Trac nodded his head and broke out into the gayest, most innocent and infectious laughter. At once the three of us were laughing together. Nothing more was said that evening, but the next morning I said quite seriously, You must never make that cake again, we didn't care for it. All Trac said, but with a wide smile, was, Me know, me know.

Of course there was no way of knowing how Trac prepared any of his delicious food. He was not secretive, but he was master in his kitchen. Much later, when he had left us and returned to us twice and then married, his wife told me the ingredients he used in some of the dishes he had cooked for us, but even she never knew the measurements. Trac said he didn't measure.

Trac left us for the first time because he was restless and wanted a change. He went off saying he would return to us and would bring us gifts. That year with Trac had spoiled us. I suggested that he find another Indo-Chinese for us. In his pretty childish way he said we wouldn't like any other Indo-Chinese, none of them were nice like he was.

And they weren't. We soon discovered that they had none of his amiable weaknesses. We had a succession of them. Each one in turn was either a gambler, which made him morose when he lost (and he always lost, for he did not work when he won), or he drank, which was unthinkable in our little home, or he loved women and would become dishonest, or he was a drug addict and he would not be able to work. Of the many we tried before Trac reappeared, Nguyen was the most satisfactory and one of the three best cooks we were to have in our long and varied experience. He would drink gently and harmlessly, for he cooked marvellously. He had been a servant in the household of the French Governor-General of Indo-China, who brought him to France. He was not young when he came to us. We took him to Bilignin for the long summer vacation, where it was obvious that he could not cook as elaborately as he did and do anything else, so we agreed that he should send for a friend of his in Paris to do the household work. Gertrude Stein and I drove over one dark night to meet the friend at the station. The new boy was well mannered. The next morning Nguyen and he were arguing in the kitchen. Nguyen said his friend had a difficult character. He seemed to be unfamiliar with household work. They quarrelled. After three days Nguyen declared that he did not suit and that we—Nguyen and I—would have to send him back to Paris. So there was a conversation. The result was that Nguyen was again contented but overworked. The Widow Roux came every day thereafter instead of twice a week. She and Nguyen bore with each other as best they could. They were as unlike as our little Chihuahua and our big white poodle. There was no love lost between either of these couples.

Nguyen cooked Chinese dishes and French dishes, and to perfection, but objected to preparing a menu with both. It was his correct sense of balance that influenced him. Both our French and American friends disagreed with him, they considered a whole Chinese menu excessive. Finally he compromised. The first course—soup, fish or shellfish with noodles or rice—would be Chinese. What followed would be French. It was suspiciously a plot to enhance the quality of Chinese cooking. In the course of time Nguyen confessed that by its delicacy and unblended flavour Chinese cooking could be remembered, and French cooking following it could not. Gertrude Stein and I thought Nguyen delightfully Chinese.

The hamlet of Bilignin was intrigued with an Asiatic in its midst and the farmers were too welcoming. They gave their guest too much wine and he got tipsy. It had become a problem in which way one should approach the subject with Nguyen. The difficulty could not be mentioned. One day to help me he said it—his health—should not be a worry to me, it was steadily improving, very soon he would be well again. And with that we dropped the question. He produced his marvellous cooking all through the summer. As with all Chinese cooks, his movements were more rapid than the eye could follow. In the afternoon when he saw the cars of unexpected friends drive into the court he would have within an hour and a half trays of the most elaborate little and big cakes and iced drinks of all kinds, as well as a *sorbet.* It was not from Nguyen that I learned, but this must have been the way he made

FRUIT SORBET

One quart *purée* of fruit, 1 cup icing sugar, put in tray of refrigerator to freeze. When it commences to harden mix thoroughly with a fork. It is unnecessary to beat it. Twenty minutes later, mix again with a fork. In 1 1/4 hours it will be stiff enough to put with a large spoon in a chilled crystal bowl and decorate with 1 pint whipped cream pressed through a pastry tube.

Nguyen was forethoughtful. He would put in the refrigerator every day a quart or more of *purée* or juice of some fruit that I had gathered in the garden and brought to him each morning. And he would see that there was a provision of whipped cream. During the freezing he would prepare his own mixture of almond paste, eggs, syrups and puff paste that he kept in reserve, and bake his little cakes. He was inventive, deft, a wizard.

This is what he would do in haste, but when he had time he made some extremely tasty combinations. For example,

COUPE GRIMALDI

Fresh pineapple cut in inch squares is macerated in kirsch for 1 hour, drained, placed in a glass and covered with mandarin orange or tangerine *sorbet*, decorated with sweetened but unflavoured whipped cream and crystallised violets.

Or a

COUPE AMBASSADRICE

Fill 3/4 of the glass with raspberry ice cream, on which place half a peach which has been previously poached in a thick syrup (1 cup sugar to 1 cup water), cooled and put in the refrigerator. Fill each half peach with 1 tablespoon *purée* of strawberry flavoured with kirsch. Surround the peach with a border of sweetened unflavoured whipped cream, sprinkle chopped skinned pistachio nuts on the cream.

Those were the names he gave these desserts, but are they or were they his inventions.

In autumn when we were preparing to close the house Nguyen went to all gatherings of the grapes for the vintage. He merited any distraction the countryside offered. It was however a pleasure for which we all paid. He came back every evening a little merrier. It was not a surprise but very disturbing. Gertrude Stein and I and the dogs had come down in the car, Nguyen by train. We had promised him that we would take him back in the car. But could we? Luckily he was soon able to tell me his health was again improving and he would be able to support the voyage. Would we. We took a chance and Nguyen was not only a perfect cook but a kind and attentive valet. Spending the night at Mâcon, he had a long conversation there with the *chef.*

Paris was however too tempting and Nguyen was no longer a possible servant for us. We parted on excellent terms, and Gertrude Stein and I remembered that Trac had warned us against his compatriots. We wondered what had become of Trac, and engaged a young Polish-American woman. She was a contrast in too many respects to Nguyen, whose cooking we would always regret. Agnel's cooking of American dishes was poor and careless; she took great pains in cooking French dishes very badly, but her Polish cooking was first rate. It was however too heavy and too rich for a daily diet. She made a Polish, not a Russian, Borshch, Polish and not Russian Piroshke. But what we liked most was

FRIED AND ROASTED BREADED CHICKEN

The chickens weigh more than 1 1/2 lbs. They are cut in four pieces—down the back and breast and across. For 2 chickens, put 1/2 lb. butter in a frying pan, flour each piece thoroughly. Beat 2 eggs in a hollow dish, coat floured chicken thoroughly with eggs, and then cover thickly with dried breadcrumbs. It is important that the chicken is thoroughly covered with flour, then with egg, then with breadcrumbs. Place in melted butter in frying pan over medium heat and brown lightly on both sides. Place in fireproof earthenware dish, wings on one side and legs on the other, backs in the centre. Put in preheated 300° oven and roast for 2 hours, basting frequently with 1 cup sour cream.

People say that they are the best Southern fried chicken they ever ate. Agnel could give the Polish names for the Polish dishes she cooked but could not write them. She was illiterate in both languages. Her mother taught her to make

VEAL AND PORK MEAT LOAF

Grind twice through the meat chopper 3/4 lb. veal and 3/4 lb. lean pork. In a bowl thoroughly mix the two meats with 1 cup chopped mushrooms, 1/4 cup fresh breadcrumbs soaked in white wine and pressed dry, 1 egg, 1 teaspoon salt, 1/4 teaspoon pepper, 1 crushed clove of garlic, a good pinch of nutmeg and 4 tablespoons sour cream. When they are well mixed form into oblong loaf, moulded around 2 hard-boiled eggs placed tip to tip. Put 4 tablespoons sour cream in a fireproof earthenware dish and place the meat loaf with enough space on all sides to be able to baste. Pour 2 tablespoons sour cream over the meat loaf so that top and sides are well covered. With the tines of a fork dipped into sour cream, press lightly on the top of the meat loaf to make a design. Bake in a preheated 400° oven for 3/4 hour basting frequently. If necessary add a little more sour cream. When done add 1 1/2 cups boiling *bouillon* to the bottom of the dish and scrape whatever may be adhering to the bottom and the sides.

At the time Agnel was working for us there were in the markets in Paris the largest frozen crawfish I ever saw. When Agnel learned that they came from Poland, we were frequently served them in a *Béchamel* sauce made with sour cream, to which a pinch of nutmeg and a very small pinch of marjoram had been added. She also prepared them by browning them lightly in lard, and pouring melted sour cream with dried brown breadcrumbs over them. Another way she cooked them was in one cup of *bouillon* of veal and 3 tablespoons *purée* of tomato.

Then unexpectedly Trac dropped in upon us and said he was ready to come back to work for us, and he would commence the following morning. We explained the delicate situation he was putting us in. He left, saying he would commence to work the following morning. Agnel was overjoyed. She would take a month's vacation on the Côte d'Azur and would find a situation there. We paid her vacation and wages and were able to introduce Trac to her the following morning. They thanked each other. It was ludicrous.

Trac had been to Indo-China as a cook on a boat from Marseilles, and he had seen his family. When he became restless there, he came back on a boat again. The parrots and a monkey he had brought back with him, and the bolts of silk he had bought in Hanoi, had been sold in Marseilles at a large profit. He would work for us for a while and then he would open a restaurant. It wasn't a secure future for any of us if Trac had these dreams, but it was extremely comfortable for as long as it lasted. Then he told us about Lucienne, his Bretonne, and their plans. He had returned to Paris to work for us and to see Lucienne. So Trac had more than one dream. He had also brought back with him Chinese dried mushrooms, peppers and powders of a kind not purchasable in Paris, so the food had new and more subtle flavours than before. One day Nguyen came to see us or him—we never could decide which. It was most unfortunate. He said he had come to tell us how healthy he now was, he had entirely recovered from the illness he had had when he was in Bilignin and he would like to work for us again. Whereupon Trac burst into an avalanche of Chinese and Nguyen left.

Trac had in some inexplicable way learned to make desserts and to bake cakes during his absence, perhaps on the boat. It was an accomplishment of which he was very proud. He said it would be useful when he opened his restaurant.

Trac had never cooked Sunday dinner for us, he left his work after lunch and returned to it on Monday morning. One day he asked me if I would teach him how to prepare Lobster Newburg. Why did he want to know how to cook that dish, I asked. He explained that since his return he had been cooking dinner on Sundays at the home of a rich bachelor who frequently gave dinner parties and whose cook did not work on Sunday evenings. Was she too perhaps cooking for someone else Sunday evenings, and so on *ad infinitum* until eventually Trac would come back to cook for us on Sunday evenings. Telling Trac how to cook something would be futile, so I told him we would cook it some day together. No, he said, it must be today, Saturday, because the gentleman said he was to cook it for Sunday evening. It seemed to me a useless effort if Trac was going to cook it with unmeasured and unwritten ingredients. But the gentleman and his guests were so pleased with the way it had succeeded that he had been sent for and had been given a glass of champagne and told that he was a great cook. Then Trac bought a *chef's* white starched cap.

This may have helped him to decide to marry Lucienne at once and open a restaurant. He said to do all this he would have to leave us soon. He brought Lucienne to see us—she was a very good-looking, educated, middle-class Bretonne. She was in love with Trac and called him her little Jean—she was making a Frenchman of him. He was happy and flattered. And in a month they had married, found a suitably small restaurant to rent in the quarter, redecorated and refurnished it attractively, and were installed and waiting for customers. We were naturally among their first. As they were expecting us, Trac made a special effort. The menu consisted of many dishes, proper to Chinese cooking. Lucienne and Trac were very happy to be restaurant keepers, but we were sadly embarrassed to find that our table was the only one occupied. Lucienne as cashier sat in the French manner high at her desk near the door. We persuaded her to eat lunch with us at our table where Trac joined us when he could spare a moment from the kitchen. He was wearing his *chef's* tall cap. When we were leaving, they said we would bring them good luck. We could highly recommend the little restaurant to our friends, and they were indeed pleased with the food, but alas reported that there were few if any customers.

When once more it became necessary to find a servant, it was more difficult than ever. But one morning, in one of the American newspapers, in Paris, there was an advertisement of what must have been the servant we were hunting. When she came to see me, she would unquestionably not find the situation with us the one she was hunting. She was a pale but vigorous, cultivated, smartly dressed Finn. She had all the qualifications to please us, and to my astonishment was ready to work for us. After telling her the disadvantages and inconveniences she would encounter, she still was ready. To be sure her wages were very high, more than double the usual wages. It was all arranged that she should come when Gertrude Stein came into the studio and was told what had been decided. Do you know if she can make desserts, she said to me. To which the Finn answered, Any dessert, Miss Stein, you may wish, from a simple Brown Betty to a *soufflé en surprise.*

So Margit was for several years the joy of our household, though not a radiant presence. She had a northern melancholy, and was called Mademoiselle Hamlet by a French friend. Her constant depression was not contagious—it became a game to combat it. There were several privileges she had with us; she had realised these before she agreed to come. As soon as she had seen Gertrude Stein she had recognised her. Margit was an omnivorous reader—there would be a library in which she would find what she longed for. She saw Gertrude Stein's easy democratic approachableness—there would be conversations with her. She would be with us and our friends in a sympathetic atmosphere. We got on excellent terms with each other. She said to us one day that she supposed there was no objection to her borrowing autographed volumes if they were not taken to the kitchen.

As a cook she was neither melancholy nor intellectual, but perfect. Margit could cook anything you asked, but she would not (indeed she refused) draw up the menus, for some unexplained undiscoverable reason. It was the only thing she expected of me. It was little enough to do in return for the excellence of the way she prepared them. For example, she frequently made for ourselves and our guests a

COVERED COCK WITH CUMIN

This is a classic of the French kitchen. It is not however a cock but a fine large chicken that is required. Boil 1 teaspoon powdered cumin in 1 cup olive oil for 1/4 hour and then put aside. Prepare the dough to cover the chicken by mixing 2 cups creamed butter, an egg slightly beaten, 1/2 cup cold water, 1/2 teaspoon salt and 1/4 teaspoon pepper. When this is well mixed work into it with a blender 4 cups flour. Place on a lightly floured board and knead with the palm of the hand into a large sheet. Gather together and repeat this operation. Roll into a ball, wrap in waxed paper and put aside in a cool place for 3 or more hours.

For the dressing, chop the liver of the chicken, 1/4 lb. calf's liver, a calf's brain previously soaked in water for 1 hour and then boiled for 10 minutes, 1 truffle, 3 shallots and 1 handful of parsley. Put this into frying pan over medium heat with half the cumin-flavoured oil and a brandy-glass of good brandy, 1/2 teaspoon salt, 1/4 teaspoon pepper. Take from fire and add 1 egg. Mix well and put into the cavity of the bird, sew or skewer carefully. Completely cover the chicken with large thin slices of pork back fat. Roll out the crust into two large rounds which will cover the chicken. Place the chicken on one, the other on top, thoroughly pinch the two together all around. In the centre of the top crust carefully cut a round opening of about 1 inch diameter to permit the steam to escape. Wrap the chicken and its covering in waxed paper as hermetically closed as possible. In a fireproof earthenware dish pour the rest of the cumin-flavoured oil. Place the chicken in the dish and roast in preheated 375° oven. Roast for 1 hour more or less, according to the size of the chicken, allowing 1/4 hour more than usual because of the crust. Baste frequently. Twenty minutes before it is done, remove waxed paper to brown the crust. This is a delicious variant of the usual roast chicken.

Margit was not inventive. She was forethoughtful, rose to emergencies and met all unexpected situations with calm. Her one ambition was to get to America, and the only obstacle was the quota, for applications had already been filed for the next seven years. Margit would frequently say that Gertrude Stein should use her influence to see that she did not have to wait so long. She did not understand that such influence as Gertrude Stein had was not in official quarters.

While we were flourishing so happily under Margit's efforts Lucienne and Trac had not been so fortunate. Their restaurant in spite of its attractiveness and Trac's good cooking had not succeeded as Trac had counted on. It was not a quarter where appreciation of Chinese cooking would be found. Lucienne said the only customers who came for meals were Trac's Chinese friends and they did not pay. They were forced to sell the furniture and the good will of the restaurant to pay their debts. They had lost all their savings. There was no pleasure in having prophesied that this would happen. Lucienne bravely decided that they would go to work together as servants. She had not been in service before, but for several reasons felt this the safest course. They would use our enthusiastic recommendation. In a few days Trac came to say that our friend Madame de C.T. was then ready to engage them. He was very excited, nothing could have been more fortunate. They would learn everything about being servants from a person accustomed to fashionable society as Madame de C.T. was. She would tell them what she expected them to do and be exacting in the way they did it. Trac was not pleased when he was told that she would expect them to know such ways and would not expect or undertake to tell them what or how to do anything. He glossed over this and said she would appreciate his cooking. She had asked him if he was able to cook a number of dishes. As he had not understood he had replied yes to all the questions.

There was a telephone call from his future mistress, and the reference given was as warm as it could possibly be. They were dismissed as inadequate after two weeks, but with Trac's good luck they found some wealthy internationals with whom they stayed for many years. Frequent absences of their master and mistress made it possible for Trac and Lucienne to satisfy their wandering habits.

Margit continued to cook excellently a variety of food. She prepared home-made

CANNELONI

Mix thoroughly 5 eggs with 3 1/4 cups flour, a pinch of salt. When this is quite smooth add 2 tablespoons tepid water. Place on floured board and knead with the palm of the hand into a long strip, roll into a ball. Repeat this operation. Cover with waxed paper and put aside for 1 hour.

The filling is made by chopping 2 cups white meat of chicken, 2 cups chicken livers, 1/2 cup ham, 1 cup mushrooms. Mix this well, add 1/2 teaspoon salt, 1/4 teaspoon pepper and 2 eggs. Mix thoroughly.

The sauce is made with 4 tablespoons butter melted in saucepan over low heat. Add 3 tablespoons flour, stir with wooden spoon, and when it commences to bubble add very slowly, stirring constantly, 2 cups hot chicken or veal *bouillon.* Slowly bring to a boil. Add 1/2 teaspoon salt, 1/4 teaspoon pepper. Simmer for 1/2 hour, stirring so that no lumps will form. Add 2 cups cream. Just before pouring over canneloni, mix in 2 yolks of eggs, heat but do not boil.

Roll out the dough to 1/10 inch thickness, cut into squares of about 7 inches. Put them gently into 3 quarts violently boiling salted water. Be careful they do not sink to the bottom of the saucepan. With a fork, prevent this and their sticking together. As soon as the water comes to a boil again, reduce heat and boil gently for exactly 7 minutes. Take from heat, drain and pour cold water over them—to put them under a tap with a spray is the quickest and best method. Drain, place on clean dishcloth and wipe dry, but really completely dry. Spread out the canneloni on a board, fill with filling at one end, roll up and pinch ends. Place on serving dish that has been liberally buttered. Do not place one on top of another. Pour the sauce over them, sprinkle 1 cup dried breadcrumbs over canneloni, also 1 cup grated Parmesan cheese, and 1/4 cup melted butter. Brown in 400° oven for 1 hour.

This is of course not an Italian way of making canneloni. It is a Finnish way, most likely. Margit made them with lobsters and hard-boiled eggs, for which she used half sherry and half chicken *bouillon* for the sauce. With asparagus tips, green peas and string beans, she used a tomato sauce.

Margit went with us to Bilignin where to our surprise she took a lively interest in the families of the farmers far and near. She would make decorated cakes of all kinds for the various holidays they celebrated, and she for them was an affable and distinguished foreigner. She was pleased and flattered when one of the young farmers wanted to marry her.

Books were sent to Gertrude Stein by some of the young writers which shocked Margit; she hoped that these were not the books young American girls were reading. In preparation for the fulfillment of her hope of getting to the United States, she occasionally cooked American dishes for us, with her usual success. Our friends in the Bugey were intrigued and amused by our succession of good cooks. Margit enjoyed their critical appreciation. She made for them a

TRICOLOURED OMELETTE

Make a *purée* of spinach by placing 1 lb. of well-washed spinach in a saucepan without water over medium heat. Turn constantly from the bottom so that it does not scorch, until it boils. Then turn flame to lower heat and cover. Cook for 10 minutes. Remove from flame, drain off water, place under running cold water until cold. Drain and press out as much water as possible. Return to saucepan over very low flame and dry out completely. Then press through fine sieve with potato masher. Mix this with 4 eggs and 1/2 teaspoon salt. Make an omelette cooked only enough to fold. Put aside and keep hot. Make another omelette of 3 yolks of eggs and 4 whole eggs, 1/2 teaspoon salt and 1/4 teaspoon powdered saffron. Place the spinach omelette on this and fold the saffron omelette over it. Place on the serving dish with a tomato sauce made by heating 4 tablespoons *purée* of tomatoes with 2 cups dry white wine. When hot pour slowly into 4 tablespoons melted butter that has been mixed over medium heat with 1 tablespoon flour. Stir until smooth and about to boil. Then add 1/2 teaspoon salt, 1/4 teaspoon pepper, a pinch of cayenne, of cloves and of nutmeg, and 1 tablespoon onion juice. Allow to simmer for 1/4 hour and add 4 tablespoons butter. Do not allow to boil. Pour around the 2 omelettes and serve. This is an effective and tasty *entrée.*

Margit prepared for our guests complicated, delicious and frozen desserts, but the favourite was

FARINA PUDDING

Put in a saucepan 1 quart milk, 4 tablespoons sugar and a pinch of salt. When it boils sprinkle into it 3 tablespoons farina (a granulated cereal). Cook for 20 minutes. Remove from flame and add 1 tablespoon orange-flower water and 2 beaten eggs. Stir from time to time.

Put 1 lb. dates and mash through a sieve with a potato masher. Mix with 1 cup unsalted butter until completely amalgamated. Add the milk-farina-egg mixture and pour into well-buttered mould. Place mould in pan of water and bake in 350° oven. Do not allow the water to boil. When a knife put in the centre of the pudding comes out dry, remove pudding from oven and pan of water. Do not turn out until tepid. Lightly coat with honey the tops and sides. Decorate top with thin strips of angelica. Serve surrounded by 3 cups vanilla ice cream and 2 cups whipped cream flavoured with 2 tablespoons kirsch.

When we returned to Paris, Margit, who had always read the newspaper avidly and who was passionately interested in international politics, was worried. She said she did not know what her future plans were. She disturbed us by her preoccupation with what was obviously not her affair. Her cooking became careless as she became more and more absorbed in some difficulty we could not fathom. Suddenly one day she said she must leave and at once, she was being followed by the police. She would return to Finland where she would be safe. She bought her ticket, had her passport visaed, and said goodbye to us that afternoon. We never knew whether she had been in real trouble or if she had imagined the whole story. She sent us a postcard from Helsinki and that was the last of Margit.

After that and until the vacation of 1939 we had to accept the services of several temporary servants of whom nothing was memorable. The summer commenced more happily than it was to end, with the good Widow Roux promoted to the kitchen and a nice country girl for the housework. With the declaration of war we, like everyone else, adapted ourselves as best we could to the new conditions. The old life with servants was finished and over.

XI.

FOOD IN THE BUGEY DURING THE OCCUPATION

IN THE BEGINNING, LIKE CAMELS, WE LIVED ON OUR PAST. WE HAD BEEN well nourished. The Bugey is famous for its food and we didn't feel hungry until some weeks after strict rationing had been enforced. The meat allowance of a quarter of a pound a week per person was not altogether satisfying, but until the Occupation powers forbade fishing, the Rhône nearby supplied us with salmon trout and the Lac-de-Bourget with the rare salmon carp, *ombre chevalier*, lavaret and perch. From the vegetable garden we had quantities of all kinds of vegetables and fruits of an excellent quality, in the wine cellars a delicious dry white wine. We were really very well off. What was lacking was milk, butter and eggs. There was an infinitesimal amount of these on our ration cards, but by the time the Germans had collected their requisitioning there was nothing left to distribute to the inhabitants. The German soldiers were interested in butter. It appeared that many of them had never tasted it. Had not Hitler asked them if they wanted butter or guns and had they not given the right answer? One day, marketing for whatever unrationed food might still be for sale, a German soldier came into the shop. He pointed to a huge mound of butter and said, One kilo. One kilo, the clerk exclaimed. The German nodded his head impatiently. The butter was weighed and wrapped up. Unwrapping one end of the package the German walked out of the shop. From the open door where I was standing I saw him bite off a piece of the butter. It evidently was not what he expected it to be for with a brusque movement he threw it violently over the garden wall of the house opposite. The story got about. People came to look at it. No one would touch it. There it stayed.

The farmers about us would not sell their produce. They would barter it for coffee, sugar, men's boots and shirts and women's smocks.

We of course had none of these things up in the attic as all French families had. One day the Germans forbade fishing so we went into Belley to see our nice butcher. He explained that the Germans rationed him to the amount of his clients' coupons. But he could regularly get us crawfish. They were caught higher up in the Valromey, preferably in an open umbrella with bait attached to the end of the ribs. That is the way we had been taught to fish them. The next morning the nice butcher came out on his bicycle to Bilignin with a sack on his back, two hundred crawfish which he emptied into the trough and covered with planks to protect them against the sun. He gave me a few scraps of meat with which to feed them. The sum I paid was astronomical. Three days later he came out with an equal number. We would commence to give lunch parties, and the *pièce-de-résistance* would be

SWIMMING CRAWFISH

For 60 crawfish prepare a *bouillon* with 2 cups dry white wine, 1/2 cup cognac, 3 large carrots and one large onion cut in thin slices, 1 teaspoon salt, a pinch of cayenne and 3 chopped shallots. Boil covered for 1/2 hour. Then put in crawfish and boil for 10 minutes, turning them about three or four times. Serve hot, cold or tepid.

Our guests brought their own bread or gave me their coupons. Gertrude Stein's and my ration went to our dogs. When war was declared Gertrude Stein wheedled a military pass from the authorities to come up to Paris so that we might protect the pictures against concussion and get some papers and our passports. The pass was good for only thirty-six hours and Paris was three hundred and seventy miles away. There was little time to waste. In the flat in Paris we soon found that wall space was four times larger than floor space, so the idea of putting the pictures on the floor was abandoned. The passports were so safely put away that they were not to be found, but in hunting for them our poodle's pedigree turned up and I put it in my bag. Later the authorities gave a ration to pedigreed dogs and Basket was not too badly nourished during the years of restriction.

After the crawfish a vegetable salad was served with fruit for dessert. Swimming crawfish was the quickest and most economical way to prepare them but might easily become monotonous to some of the friends who would frequently be our guests. I then prepared a version

CRAWFISH À LA BORDELAISE

For 60 crawfish chop 3 large carrots, 2 large onions, 4 shallots, the value of 1 tablespoon tarragon and the same of parsley. Put all these in a saucepan with 1/4 lb. butter, a twig of thyme, a laurel leaf, 1 teaspoon salt, 1/2 teaspoon ground pepper. Cook over medium flame. Shell the raw crawfish and put them into the saucepan with a pinch of cayenne. Light 1/2 cup cognac and pour it in. Allow the crawfish to become red, then remove them. Add to the saucepan 3 cups dry white wine (preferably a Bordeaux). Reduce the sauce to about half. Then add the crawfish and 4 tablespoons tomato paste and cook for 10 minutes. Strain the sauce, bring to a boil but do not allow to boil, and add 1 1/2 cups butter in small pieces, stirring constantly.

Needless to say, the butter was omitted, the vegetables were cooked in 4 tablespoons olive oil, and 2 tablespoons were added at the last moment to make an unctuous sauce. Our reserve of 3 quarts olive oil was only exhausted shortly before walnut oil was purchasable on the black market. Walnut oil is delicious. Later it was replaced by hazel-nut oil which is more delicate.

The Germans disappeared after six weeks; we were in the southern zone. Requisitioning continued and before the autumn of 1940 any supplies that were not on the coupons were no longer to be found. The grocery stores were empty but before this had happened I had bought dried fruits, chicory to replace coffee, sardines, spices, corn meal and cleaning materials. The autumn harvest in the vegetable garden would largely see us through the winter with the string beans and tomatoes I had put up. This is the way I learned to put up

TOMATOES AU NATUREL

Skin 28 lbs. tomatoes, cut through in both directions, then cut each of these four pieces in half. Put in a pan over a medium flame. Add 7 ozs. salicyclic acid which can be bought at any good chemist's. Carefully mix the acid and the tomatoes. Heat thoroughly, stirring constantly, but remove from stove before boiling point is reached. Fill jars with tomatoes. When cold pour 1/2 inch of oil on top of the tomatoes so that no air enters. Cover each jar with paper. This will keep the oil clean so that it may be used later in cooking the tomatoes. A foolproof recipe.

Our vegetable garden had been the prettiest one for miles about. I was very proud of it and of what it had produced. The hard work had exhausted me. Suddenly we realised that we were hungry but it was not mentioned. It was at this time that I dreamed one night of a long silver dish floating in the air and on it were three large slices of succulent ham. That was all. It haunted me for the six months that were to pass before the blessed black market was organised.

Friends would come out to have a cup of real China tea with us. With economy the ten pounds a friend had sent us from the United States in the summer of 1939 lasted until the Liberation. Gertrude Stein had bought for me all the American cigarettes she could find. If they weren't nourishing, they certainly acted as a stimulant at this time. Hospitality consisted in two cups of tea without sugar, milk or lemon and one cigarette. One sombre afternoon I saw the good Widow Roux who was our handyman going to the portals with a pail in either hand. What have you got there, I asked. Our dishwater for the Mother Vigne's pig, she answered. Listen, I said, you tell her if she isn't ready to sell us an egg a day you won't bring her any more dishwater for her pig. The Mother Vigne accepted the proposal and our diet was appreciably increased. It was manna from heaven. With reasonable calculation, if the Mother Vigne was now selling us one egg a day, in a short while when her hens began to lay normally again she would certainly be willing to sell us two a day. It was a comforting thought, but a few days before Christmas my hopes were shattered. The Mother Vigne's son told me that he was killing the pig for the holidays. As they would no longer have any need for the dishwater his mother wanted to say that she was not selling us an egg each day. It was a blow. Perhaps something else would turn up.

The interminable winter dragged on. A friend or two ran the blockade and came down from Paris with news. But our most important news came from a friend, Hubert de R., who was in the *Résistance.* He would bicycle over from Savoie and lunch with us. The appetite of a husky man after pedalling eighteen miles in the snow on uncleared country roads had to be met as best one could. Hubert had a sweet tooth so a dessert was in order. One day I prepared for him what I called

RASPBERRY FLUMMERY

Two glasses raspberry jelly melted in a double boiler with 1 1/2 cups water to which is added the juice of 1/2 lemon and 2 or 3 sheets gelatine, depending upon their size, soaked in 1/2 cup water. Pour into a mould. When cold put in refrigerator. Turn out of mould to serve.

The flummery cried for cream. So did we.

Gelatine was rare but I had a large quantity in reserve. Hubert de R. enjoyed his dessert. Around the fire after lunch he said,

That dessert was made with gelatine, wasn't it? Where do you find gelatine these days? There is none in Savoie. My wife no longer has any. His knowing anything about gelatine surprised me. When as he was leaving I gave him twenty sheets to take to his wife, he was more grateful than the small gift justified. It was not until some time later that he told us for what he had wanted the gelatine. He had needed it desperately for making false papers.

Gertrude Stein was still allowed to run her car, transformed from gasoline burning to wood alcohol. One day we went over to take the Baronne Pierlot for a drive. She said she would like to stop at a little shop where she had heard they were selling rice on the black market. While her daughter-in-law was in the shop Madame Pierlot said, It remains to be seen what success she will have. Has she or has she not the right personality, for it is not with money that one buys on the black market but with one's personality. This was indeed true. Later Gertrude Stein when no one else did would return from a walk with an egg, a pound of white flour, a bit of butter. The cook and I would welcome her upon her return.

We heard nothing more of the black market until spring when a rumour reached us that across the valley in Artemarre the well-known *chef* B. was serving excellent food to his old clients. Gertrude Stein at once proposed that my birthday should be celebrated with a lunch party at Artemarre. We telephoned to B. and told him that we and some of our friends wanted to come over on a certain day to say how do you do to him. One was discreet on the telephone those days as one was everywhere else in public. He answered that he would be enchanted to see us all again, and as an afterthought asked how many we would be. So a dozen friends would meet us at Artemarre at one o'clock. Means of transportation were strange and varied. We would go over with the doctor and his wife in their little car—doctors had a small ration of petrol—and he would visit his patients on the way. Five of our guests drove over in their farmer's high two-wheeled trap driven by an enormous old nag the Germans had found useless at the time of the requisition. Two of our friends came in a miniature dog cart painted in all the colours of the rainbow drawn by a Shetland pony, all of which they had lately acquired for a large sum from a circus family encountered on the road. The others bicycled or walked over. In spite of the gloom of the Occupation we were delighted to meet again and to anticipate the feast B. was cooking for us. We had gone to see him in his kitchen where the unaccustomed fragrances and the menu he showed us were pleasurably exciting. This was the prodigious repast we sat down to:

Aspic de foie gras
Truites en chemise
Braised pigeons—shoestring potatoes
Baron of spring lamb—jardiniere of spring carrots—onions—asparagus tips—string beans *en barquette*
Truffle salad
Wild-strawberry tart

B. obligingly had given me many years before his recipe for

TRUITES EN CHEMISE

Each trout is slit down the belly, emptied, washed and dried, and stuffed with as many chopped mushrooms as it will hold moistened with a squeeze of lemon juice, and 1 tablespoon cream per 1/2 lb. fish seasoned with salt and pepper. The opening is then carefully skewered together. Bake in enough melted butter to float them in a 480° oven for 15 minutes, basting three or four times. The fish should be cooked through but not browned. Take from the oven and drain. Roll in a *crêpe*, folding the sides of the *crêpe* in. Place on a metal dish over boiling water covered with waxed paper. Serve hot.

CRÊPES

Sift 2 cups flour into a bowl, stirring constantly with a wooden spoon, slowly add 5 slightly beaten eggs, 1/4 cup tepid water, 1/4 cup beer, 1 tablespoon rum, 1 tablespoon kirsch and 1/4 teaspoon salt. The paste should be perfectly smooth before putting aside for 4 hours. Then stir again thoroughly. Heat small iron sheet over medium flame. Melt butter the size of a hazel nut and spread with a pastry brush. Pour 2 tablespoons paste on the sheet and cook until the underside is pale gold. Turn over with a wide wooden spatula on to the sheet which has been lightly brushed with butter. When the second side is pale gold, take off, place on a heated platter, put a cooked trout on one end and roll the *crêpe* around it. Trim off with scissors any unsightly ends. Put aside on the serving dish covered with waxed paper. Wipe the iron sheet clean and continue with each *crêpe* in turn, not forgetting to butter the sheet for each side of the *crêpe.*

It does not take much experience to become expert in making *crêpes.* It requires a light quick movement. We had a cook who tossed the *crêpes* high into the air. She liked our friends to come to the kitchen door and see her do her little trick. An Indo-Chinese cook we had did not use a spatula either. He simply turned them deftly by hand.

Before lunch was served we all talked at once. There was a great deal to tell each other of what had been happening during the many weeks since our last reunion. But as soon as lunch was served a silence of astonishment, contemplation and satisfaction descended upon us. This is typical of the French, particularly in the provinces. No one talks as the first course is served. Frank discussion of the food that is served is also characteristic of the French. It is not considered bad manners. What is, is the way we Americans do not eat all we have on our plates. This they consider unpardonable. They have, however, by taking a small portion, the opportunity of compensation from the invariable habit of a second serving. After so long a fast we were pleased to indulge in more food than was perhaps going to be good for us. Someone remarked that fasts should be broken by a glass of orange or tomato juice. Eat, drink and be merry, said I. Ah, if one were only certain in these days of dying from overeating. One remembered the packages of food one was sending to war and political prisoners and felt conscience-stricken at the overabundance of our feast. We did nevertheless recover our high spirits before we parted, having thanked B. for his sumptuous though delicate lunch and promised to return soon.

That lunch was the beginning of the excitement and gratification that came to us gradually from provisions secured on the black market. In late May our friend of the circus cart—it was now sedately painted dark blue—and pony drove over to see us. Would we share half a lamb with her. A farmer near her home was clandestinely killing two tomorrow, one for himself and one for her. She would be glad to divide hers with us. Naturally we accepted. She would make the best terms possible with the farmer. The next morning she drove over again with a sack at her feet. She, I and the good Widow Roux did our first butchering. Keeping out some chops for our lunch, we put the rest of the precious beast into the refrigerator. Would such a windfall reoccur? To our surprise it did. Not too soon to be sure for us not to have to return to lean days and even weeks, but we had mysterious visits from unknown men and women with odd bits of frequently unknown food—that is, pigs in pokes. *La Veuve* Roux would come to say there was someone in the kitchen who wished to speak to me. Or she would have a quarter of a pound of butter, a sausage, a quart of milk, sweetbreads and brains, all greatly appreciated. A sheep's head was a feast for the poodle. For some time we had strange and varied food. Then we Americans came into the war, and our landlord, an officer in the French Army, required his home and we were forced to move. We were broken-hearted to have to leave Bilignin. Friends found us a house at Culoz and we moved there the day the Germans occupying the southern zone came into Belley. At Culoz we should be less favoured. We had no acquaintances there and the country round about was less productive, only there would be more fine dry white wine. On the large property there was no vegetable garden. It would be starting over from scratch. With the house went two servants, a very fine cook who announced at once that she could not cook with the scanty materials the coupons allowed. She was not encouraged when she was told that the black market would largely supplement them. She was old, tired and pessimistic. So it fell upon me to do most of the cooking while a great cook sat by indifferent, inert and too discouraged to pay attention when I tried to show her how to make

A RESTRICTED VEAL LOAF

One cup chopped veal, 3 cups breadcrumbs soaked in a dry white wine (Glory be for the inexhaustible provision of it!), 1/4 teaspoon pepper, 1/2 teaspoon powdered basil, 1/2 teaspoon powdered tarragon, 1/2 teaspoon powdered chervil, 1 teaspoon powdered parsley, 1/4 teaspoon powdered bay leaf, 2 chopped onions, 3 chopped shallots and 1 treasured egg. Mix thoroughly, form into a loaf with a greased knife and place in greased earthenware dish. Bake in 375° oven for an hour basting with dry white wine.

The herbs were from the vegetable garden at Bilignin, dried and powdered and kept in airtight jars. They suggested a meat flavour. Their general use at this time, and not only in France, is I think the reason that herbs are so popular today.

HOME-MADE MUSTARD

Pound in a mortar 1 tablespoon chopped parsley, 1 tablespoon chopped tarragon, 1 teaspoon chopped chervil and 1/2 cup mustard seed. When reduced to a powder strain through a fine sieve and slowly add 1 tablespoon oil and 1 tablespoon vinegar. Stir thoroughly and keep covered. In season the strained juice of

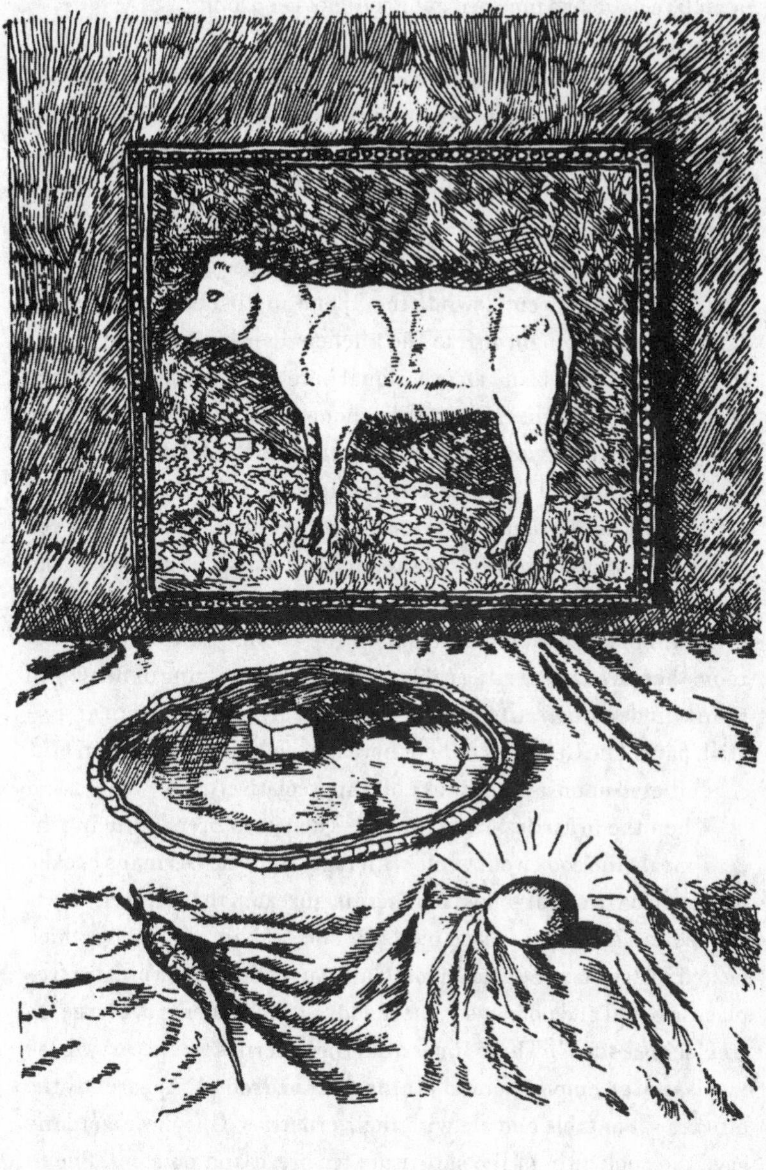

gooseberries or currants may be substituted for the vinegar. These make a delicious mustard but it will not keep long.

The mustard seeds had been a chance purchase one day when there was nothing else to be found. Two pounds of them permitted all our friends to partake of a relish that had disappeared long before.

In the village two of the shopkeepers were to become very useful to me. They said it was their patriotic duty to sell what the Germans forbade. In which case was it not mine to purchase what they offered? The country boys went down to the Rhône and fished clandestinely. They not only brought fish to the kitchen door but flour, lard, nuts in small quantities, and an occasional hare or rabbit. We gave lunch parties. The cook cheered up a bit, though she grumbled that there was nothing to cook anything with but the inevitable white wine.

Suddenly we had Germans billeted upon us, two officers and their orderlies. Hastily rooms were prepared for them in a wing of the house far removed from our bedrooms. Provisions were hidden, but there was not enough time to gather together and put away the many English books scattered throughout the rooms. In the best guest room there was a charming coloured English engraving of Benjamin Franklin demonstrating one of his discoveries on a lake in an English park. The Germans did not notice it, but one of the Italian officers billeted upon us later spoke of it appreciatively.

When the orderlies came into the kitchen to prepare their officers' meals the cook went white with rage. How the Germans cooked has no place in a cook-book, but their menu eaten three times a day is offered as a curiosity. Per man: 1 large slice of ham 1 1/2 inches thick heated in deep fat, the gelatinous-glutinous contents of a pint tin (replacing bread and potatoes?), the muddy liquid contents of a large tin (replacing coffee?). Three times a day the orderlies would carry these meals into an empty room adjoining their bedrooms. Apparently the officers sat at table and ate with their orderlies. One day an orderly gave the cook a tin of the substitute for bread and potatoes. She in turn gave it to our most treasured possessions, four hens. They ran eagerly toward it, pecked at it and walked away. The cook, delighting in the *geste* of her French hens, threw the mess into the mountain torrent that ran around two sides of the house.

After two weeks, the Germans billeted upon us left. We gave a deep sigh of relief.

The Germans had requisitioned the automobiles in Culoz, including the two taxis, but occasionally when the Mayor had to go over to Belley or someone had to go into hospital there Gertrude Stein and I would be tucked away in the car. And once in Belley we would have an orgy of seeing our friends and of running in provisions. The cake shop and confectioner was famous not only throughout the Bugey but throughout France, particularly for

TRUFFLES DE CHAMBÉRY

Melt 1/4 lb. chocolate over boiling water, add 2 tablespoons butter and 1 1/2 tablespoons powdered sugar. Stir until sugar is melted. Remove from boiling water and add the yolks of 2 eggs one at a time, stirring constantly. Add 2 teaspoons rum and mix thoroughly. Put away in a cool place (not the refrigerator) for 12 hours. Then shape into small balls and roll in powdered chocolate. This makes a very small quantity. They are exquisite.

Madame Peycru ran the store, her husband stayed in the kitchen. She had cakes under the counters—what was shown was for the Occupation Forces—not good enough for you, she said to me as she served them. She would gaily pass me a package with an outer wrapping of newspaper.

Finally the Germans relinquished two broken-down buses which made a daily trip from Culoz to Belley and return. It was then that Madame Peycru proposed to send us cakes. I went to the bus stop to retrieve the first one. Intending to be discreet, she had addressed it to The Two American Ladies in Culoz. Not one of the two hundred and fifty Germans and their officers stationed at Culoz suspected our nationality, the French authorities having destroyed our papers and done everything possible to protect us. There was nothing to do but to hope for the best and to take the bus over to Belley to warn Madame Peycru neither to put either of our names on the package nor to mention our nationality. The conductor of the bus would leave it at the café where one of us would pick it up.

As the dreary dismal months dragged on provisioning became easier and more abundant, except for meat and butter. More people came to see us, even from Lyon, which is seventy miles distant. All in the *Résistance*, naturally. They brought chicory to replace the bird seed our coupons provided as coffee. In return there were eggs for them, and if warned in advance the baker at Belley would send over a

BRIOCHE

Sift 2 cups flour into a bowl, add 3 eggs, stir with a wooden spoon until completely smooth. Continuing to stir add 1 more egg, 4 tablespoons milk in which 1 cup sugar has been mixed with 1 tablespoon rum and 3 teaspoons baking powder dissolved in 1 tablespoon water. Mix thoroughly. Add 3/4 cup softened butter. Mix well, put aside for 12 hours in a cool spot (not in the refrigerator). Pour into deep round buttered mould and bake for about 30 minutes in 375° oven. These *brioches* are very nice baked in very small muffin tins, for about 20 minutes in 350° oven.

Little by little we had a greater abundance of food. We acknowledged that we were no longer hungry, but there was a hopeless monotony in the menus. Fish was our most substantial and nourishing food. It was a protracted, indeed a perpetual, Lent. After a raid on Belley some exceptional delicacy might be the spoil. A few slices of sausage or an infinitesimal piece of cheese could only add a faint flavour to a dish. It was then that I betook myself to the passionate reading of elaborate recipes in very large cook-books. Through the long winter evenings close to the inadequate fire the recipes for food that there was no possibility of realising held me fascinated—forgetful of restrictions, even occasionally of the Occupation, of the black cloud over and about one, of a possible danger one refused to face. The great French *chefs* and their creations were very real. Gertrude Stein had the habit of giving me for Christmas a very important cook-book—even during the Occupation she would surprise me with one. When all communication with Paris was forbidden, the 1,479 pages of Montagne's and Salle's *The Great Book of the Kitchen* passed across the line with more intelligence than is usually credited to inanimate objects. Though there was not one ingredient obtainable it was abundantly satisfying to pore over its pages, imagination being as lively as it is. This is their recipe for

TOURNEDOS MARGOT

Allow for each person a slice weighing 1/4 lb., after fat and skin have been removed, of tender loin of beef from the centre. Fry in butter on both sides over a quick fire to sear the outside but leave the centre rare. Salt to taste. Remove from the pan when done and place on each slice the caps of very large mushrooms previously cooked in butter. On these place finely cut truffles in a cream sauce. The tender loin slices and their filled mushroom caps are served on a mound of Anna potatoes.

The truffles are peeled, diced and cooked in butter very slowly in a covered saucepan for 15 minutes, shaking the pan frequently. Add 1 teaspoon best brandy for 1/4 lb. truffles, cover with boiling cream and again bring to a boil. Remove the truffles. While the cream is reducing in the uncovered saucepan add 1 tablespoon thick white sauce. When the cream has been reduced to half its quantity replace the truffles in the sauce. Allow to come to a boil. Remove from fire and stir in 1 tablespoon butter cut in small pieces. Fill the mushroom caps with the truffles and their sauce.

Anna potatoes should be firm, the kind suitable for frying. Peel and cut into 1/8-inch slices. Put overlapping in layers in very well-buttered low moulds not much wider than the fillets of beef. Pour melted clarified butter between the layers and on top of the last layer. Cover hermetically and cook in a 400° oven for 20 minutes. Turn upside-down and cook for 10 minutes longer. Remove from mould, discarding superfluous butter. Place under the fillets with their truffle-filled mushroom caps and serve at once.

One day the cook told me she had in her possession the manuscript cook-book of the grandfather of our landlady, Monsieur Lucien Tendret, a well-known judge and a famous *gourmet*, author of *Food in the Country of Brillat-Savarin.* Gertrude Stein had been given a copy of this delectable book by a grandson of the author. We used to speak of translating it into English. It is of course not a proper cook-book, as the book the cook proposed to allow me to read was. The recipes are exciting to read but are not useful even today. Take for instance

LOBSTER, BREAST OF CHICKEN AND BLACK TRUFFLE SALAD

Pour into a salad bowl the best olive oil, white-wine vinegar, 4 tablespoons juice of roasted turkey, 1/2 teaspoon tarragon mustard, the interior of a lobster, salt and pepper. Turn until perfectly smooth. Put into this the sliced lobster meat, sliced breast of a steamed skinned capon, sliced breast of a roast turkey, also skinned, and of three young partridges (keep the finest slices for decorating), thin slices of truffles previously cooked in excellent dry white wine, the same of mushrooms and a quantity of shelled crawfish. Cover with a layer of the white leaves of escarole or endive. Arrange one more layer and on the second layer of escarole tastefully arrange the reserved slices of breast, a few narrow strips of ham from which the fat has been removed, some large slices of truffles and of mushrooms, a border of shelled crawfish, 1 tablespoon capers washed in white wine, and 1 cup stoned green olives. In the centre arrange a tower of very stiff mayonnaise surmounted by the largest truffle. With this salad the best dry champagne should be served very cold but not iced.

By the autumn of 1943 food was no longer a subject. We were impatiently waiting for the invasion and liberation. In 1940 I had accumulated amongst other prized provisions 4 lbs. citron, candied orange and lemon peel, pineapple and cherries and 2 lbs. raisins, all of which I had put into two well-covered glass jars. Our friends knew of this *cache* and that it was being saved for a Liberation fruit cake. From time to time I would look at the jars to see that the fruit was not getting dry. They cheered me greatly during the increasingly dismal days of that winter and the early spring of 1944. Spring finally did come. Even a potato blight on the early plants didn't discourage us—we picked the wretched beetles off by hand. When the *Résistance* derailed the engine of a train entirely composed of huge vats of Spanish wine destined for Switzerland our Mayor, fearing he would be unable to safeguard it, requisitioned the wine. He divided it among the 250 members of the local Occupation Forces and the 1,500-odd inhabitants. We each received twelve quarts. Gertrude Stein's and my share were put aside for Liberation festivities. The servants, no mean connoisseurs, reported that it was indeed very good.

Just when the communiqués were getting almost unbearably exciting, two officers and thirty soldiers of the Italian army were billeted upon us, the officers in the house naturally and the soldiers in the garages and chauffeur's quarters—worrisomely near the vegetable garden and fruit trees. Would they respect what was missing from their army rations? Their captain said they would and surprisingly they did. Occasionally one of them would come to talk to me while I was working among the vegetables, the wind wafting savoury and tantalizing odours of *minestrone* from a huge pot cooking under the trees below. Presently the soldiers were selling me on the black market such cigarettes as they could spare, a most welcome relief from my tobaccoless state. I had been smoking something called garden tobacco. The Germans gave a limited ration of tobacco plants to men, none to women. There were no proper facilities for drying or for cutting the leaves, but we smoked anything we could roll except fig leaves, which had poisoned a friend. The Italian tobacco was agreeable, convenient and plentiful; our young maid found the soldiers equally so. The two officers billeted upon us sent over three pounds of Parmesan cheese. We were overcome by the sight of it and invited everyone to a party to eat (an abbreviated)

FONDUE

Wash and peel 5 ozs. truffles cut in small dice. Cook in 2 ozs. butter stirring constantly. Put aside. Separate whites from yolks of 12 eggs and strain each into separate bowls. Stir the yolks thoroughly. Beat the whites until they no longer pour from the bowl. Add the yolks slowly to the whites. Salt and pepper. Add 1/4 lb. butter cut in small pieces, 1/4 lb. grated cheese and truffles. Put 1/2 cup good stock into a deep porcelain or Pyrex saucepan. When it boils, pour into it the egg-cheese-truffle mixture. Beat with a wire whisk. When it commences to thicken remove from flame continuing to beat until the cream is perfectly smooth. Then add 1/2 cup more stock. Cut 1/2 lb. butter into small pieces. Replace on a very low flame continuing to whisk. It should be absolutely smooth. Add 1 small glass best kirsch. Serve on hot plates at once.

Only a quarter of the butter required was available, which made the *fondue* less unctuous than it should have been, but even so it was quite a feast to taste a *fondue* again.

The Italians stayed until their country accepted the Armistice. When they heard the news, they tore up their military papers and left singing. There were about six hundred Italian soldiers in the neighbourhood and the frontier was only 125 kilometres away. We hoped they would cross it safely. Later we heard that they had all been killed by the Germans.

Events were precipitating themselves in the happiest confusion. The northern landing had taken place. The Germans knew we knew. Discretion was thrown to the winds. We thought we heard someone singing the *Marseillaise.* That afternoon we were informed that over a hundred Germans were to be billeted upon us. Almost immediately they were at the gates, five officers and seven noncoms to be lodged in the house, the men on the terraces and in the gardens. Gertrude Stein with her manuscript and the poodle were whisked upstairs to the bedroom. The servants and I brought down the mattresses for the noncoms and prepared the beds and rooms for the officers, whose dogs were roving about the house as their nags and donkeys were roaming in the flower beds. It was a hideous confusion. That night the Germans ate cold army rations. The next morning they killed a calf on the terrace nearest the house and cooked it on an improvised spit. In the afternoon they left after helping themselves to our small supplies and souvenirs. The jars of candied fruits were safely hidden in the linen closet. That meant a lot to me—they were a symbol of the happier days soon to come.

Within six weeks after the Germans left the southern landing had taken place. We were jubilant. Our army or at any rate part of it would pass us on the way north.

The *Résistance* had not only disrupted the railroad tracks but had blown up the main highway in several places. Gertrude Stein returned from the village one afternoon excitedly announcing that the Germans had placed machine guns all over the square and on the four roads leading from it and that soldiers were coming into the village. Well, said I, we won't sleep tonight. It will all be over by morning. But nothing happened. In the morning the guns had disappeared. When we heard over the radio that Paris was liberated we were wild with excitement. The end was near. So the boys of the *Résistance* came down quietly from their mountain top one morning, drove the seven hundred Germans from Culoz and the neighbourhood into the marshes, surrounded them and wiped them out. It was glorious, classic, almost Biblical. We celebrated by taking one of the liberated taxis to Belley. Home-made flags were flying from windows—not only the tricolour but the stars and stripes, the number of stars and stripes varying according to the amount of dye procurable. Our friends were as excited as we were. There was a rumour that the First French Army and the Seventh American were at Grenoble, only forty-five miles away. Anything could happen now to the degree that when I thought I saw a jeep stop at the kerb I believed it was, and it was. It was the first jeep I'd ever seen, but it surely was one for there were undeniably two American soldiers in it. We flew over to them. They wanted us to put them on their road. We at once requisitioned them for dinner and the night—as they would have to dine and sleep somewhere, why not at our house. We had a triumphant entry at Culoz and at the house. The servants cried and curtsied and hailed them as *nos Liberateurs.* Impossible to calm the cook. Finally she agreed to go back to the kitchen to prepare a dinner, the first in four years, saying, Do not worry, madam, now I can cook even if there is no cream and not enough butter and eggs. This was her menu:

Trout in aspic
Chicken à l'estragon
Tomato and lettuce salad
Chocolate soufflé
Wild strawberries
Black coffee

For the liberators neighbours had been willing to sell her a chicken, eggs and butter. Our American colonel and his driver contributed the chocolate and coffee.

This is the way the cook prepared

CHICKEN À L'ESTRAGON

Put the neck, gizzard, liver, heart, two feet and two first joints of the wings into 2 inches more cold water than to cover. Put uncovered over very low flame and bring slowly to the boil. Skim very carefully. Do not cover until all the skim has been removed. Then add a carrot cut in small pieces, a medium-sized onion with a clove stuck in it, a small bunch of celery leaves, a stalk of parsley, 6 whole peppers, salt and a very small piece of whole nutmeg. The contents of the saucepan should be well covered with water. Cover and cook slowly for 2 hours.

Truss the chicken, brown lightly on all sides in butter in a Dutch oven over medium flame. When browned, cover, reduce flame, place water in the cover and allow the bird to sweat in its own juice for 20 minutes, then add 1/2 cup hot dry white wine, 1/2 oz. washed tarragon. Add when needed for basting a little of the giblet broth. The chicken should cook from 1 to 2 hours according to its size and age.

Before serving prepare enough washed tarragon leaves to decorate the breast of the chicken in an Indian tree pattern. It is a pleasing way to present it. Place the chicken on the serving platter. If necessary add more of the gizzard broth. Pour a few tablespoons of it into 3 well-stirred egg yolks and add this to the juice in the Dutch oven, stir carefully while heating thoroughly but do not allow it to boil. Serve in a sauce boat.

The next day two friends bicycled over from Belley to lunch with us and to taste the Spanish wine. The cook was in excellent form, she was cooking us the most fantastic food. For lunch she had suggested as dessert a puff-paste tart with a Frangipani cream. We were at table when the maid suddenly burst in crying, The American army, they are at the door. Tell them to come in, we said. They are washing their hands, she answered. Gertrude Stein and I went and welcomed them. One of them was Siberard, whom we had known in Paris, another was Frank Gervasi, for whom we felt an immediate friendliness. It was a very gay lunch indeed, though our French friends spoke no English. Siberard had come to ask Gertrude Stein to speak on the radio to the United States from the American station just installed at Voiron, which she was delighted to do. However did you know where to find her, I asked. We had a general idea, Siberard said, but we got lost. A jeep happened to come along and we asked them. The colonel seemed surprised we did not know the way. It was he who put us on the road here.

The Frangipani tart was a successful climax to the extensible lunch. The tart was a very large square one. On each corner were small French and American flags. The cook had made them of paper with two coloured pencils that she had borrowed. She had sat up until midnight making quantities of them, knowing the pencils were not going to last long.

FRANGIPANI TART

Bake a puff-paste crust in a square cake tin, lightly moisten the sides so the dough will adhere to them. It should be deep enough to hold plenty of cream. For the cream stir 4 eggs and 8 yolks, add 1 cup sugar and stir, 2 cups flour, stir until thoroughly amalgamated. Slowly add 1 1/2 quarts scalded milk. Place in double boiler. Stir until quite thick. Remove from heat and continue to stir while very slowly adding 1/2 cup butter cut in small pieces, 4 macaroons dried and rolled to a powder, 1 teaspoon orange-flower water and 2 tablespoons best Jamaica rum (this last was the invention of the cook). Shortly before serving spread evenly into completely cold puff-paste crust and sprinkle with powdered sugar.

Frangipani cream is supposed to have been created by an Italian named Frangipani who lived in Paris during the early seventeenth century.

They called for us in a jeep, in fact two jeeps, to go to Voiron where Gertrude Stein was to speak on the radio. We lunched at the Officers' Mess on almost forgotten American food—ham and eggs, tinned corn, sweet pickles, biscuits and tinned California peaches, coffee with evaporated milk—a memorable lunch it was with our liberators.

It reminded me that the long-promised fruit cake had not yet been made. So much had been happening that there had not been sufficient time to weigh and chop the necessary ingredients.

The day after the expedition to Voiron Gertrude Stein said we would go over to Aix-les-Bains without police papers to see how beautiful it must be without 8,000 Occupants. As we were walking out of the station, some American soldiers got out of a jeep and came up to Gertrude Stein to ask her for her autograph. They were evidently the first Americans in Aix-les-Bains for quite a crowd had gathered about them as she scribbled her name for them. One of them said, You live near here, don't you, Miss Stein. May we not come to see you if we ever get near here again. Then they introduced themselves, one major and three colonels. Several colonels have come to see me, Gertrude Stein said. Now I would like a general. So far no general has come to see me. One of the officers said, We are on the staff of a general, General Patch, who commands the Seventh Army. I would like him to come to see me. You tell him I would. You tell him that if he does we will give him a chicken lunch. With that we parted and promptly forgot the meeting. To our surprise, about a week after, one afternoon another colonel came with a message from General Patch to thank Gertrude Stein for her invitation to a chicken lunch which he hoped to be able to accept when the war was over, for if possible he would then be going to Aix where he had spent a happy leave during the other war. There was nothing but a very plain cake indeed to give to the colonel to take to the general, but I got the latter's military address so that I might send him

LIBERATION FRUIT CAKE

A day in advance cut in slivers 1 1/2 lbs. citron, 1 lb. cherries. Wash, dry and cover with brandy 2 lbs. white currants. Put aside. Blanch, dry and chop 1 1/2 lbs. almonds. Put these aside separately. The next day, thoroughly cream 1 lb. butter, add slowly 1 lb. sugar stirring until very light. Add the yolks of 12 eggs one by one. Drain and dry the raisins. Add to the prepared fruits. Sieve 1 lb. flour, put enough of it on the fruits to which the raisins have been added to make sure they do not stick together. Put the fruits in a sieve and remove all the superfluous flour. Add the flour in the sieve to the remaining flour, to which add 2 teaspoons cinnamon, 1 teaspoon mace, 1 teaspoon nutmeg and 1/2 teaspoon cloves. Slowly sift the flour into the butter-sugar-egg mixture, stirring thoroughly after each addition. Add the ground almonds, 1 cup best brandy and 1/4 cup rose- or orange-flower water. Beat the whites of 12 eggs and fold lightly but thoroughly into the dough. Then fold in the floured fruits and raisins. Put into buttered pan or pans, lined with buttered brown paper. This quantity makes about 12 lbs. Bake for about 4 hours according to size of pan or pans.

A fruit cake should have an inch of almond paste spread over the top before frosting the entire cake. This is not gilding the lily, it is only bringing its perfume more pronouncedly to your attention.

Almond Paste

Pound in a mortar until reduced to a paste 1/2 lb. blanched almonds that have been wiped dry. Put them in a heavy enamelled saucepan with 1 lb. sugar, 1 teaspoon orange-flower water and 1 teaspoon vanilla. Stir thoroughly over very low flame until the sugar is melted and the paste is smooth. Remove from flame, place on a marble slab or table lightly sprinkled with powdered sugar, knead in 2 tablespoons powdered sugar. Spread with a moist knife on top of the cake or cakes.

It had been possible to secure all the necessary ingredients for the Liberation Cake with the exception of the almonds. Hazel nuts and walnuts were the only nuts that grew in the Bugey. Walnuts were discarded because of their heavy flavour. The easiest way to blanch hazel nuts is to put them into the oven but that would have dried them. There was nothing to do but to remove their skins by plunging them a few at a time into boiling water—two pounds in all. It took a long time but the substitution of hazel nuts for almonds was successful. When I have time, I still use it.

With the cake sent off to the general whose army had liberated the Bugey the Occupation was over—whatever the unknown restrictions to come might be.

XII.

RECIPES FROM FRIENDS

APPETISER

Fania Marinoff
New York

HERB BUTTER

(to be spread on biscuits or crackers)

Grind or mince finely equal amounts of fresh parsley, fresh dill, 1 green pepper, raw spinach or dandelion greens, green tops of scallions, shallots or leeks, and watercress. Season with salt, pepper, paprika, garlic, thyme and tarragon (fresh, if you have it). Bind with soft butter and mayonnaise.

SOUPS

Madame M. G. Debar
La Régie
Soye-en-Septaine

MUSSEL SOUP À LA RÉGIE

Put the scrubbed and washed mussels at once into a casserole that has been lined with minced onions and garlic, put between layers of mussels other minced onions, garlic and a few thin slices of carrots. Add some good white wine. Let the mussels cook for 1/4 hour. Keep the casserole hot. Remove the juice in which the mussels have cooked. Thicken lightly with flour, add slowly to this sauce 3/4 cup cream in which the yolks of 2 eggs have been mixed. Taste and add salt if necessary—in any case add a powerful pinch of pepper!

Serve the mussels directly from the casserole, the sauce separately, as well as slices of bread browned in butter and generously rubbed with garlic.

With a Rhine wine, if possible, this dish can boast of being without reproach.

DORA MAAR
Paris

LAUREL-LEAF SOUP

Boil a branch of laurel with its leaves in a saucepan for 20 minutes. Remove the laurel. Stir 1 yolk of egg to every 2 cups of the laurel water. Add a little hot water to the yolk of egg, stir and add laurel water. Heat but do not allow to boil. Serve. *Croûtons* may be added to soup.

An invigorating soup, served during the winter in Provence.

MADAME JOSEPH DELTEIL
Grabels

IBIZA SOUP

For 4. Heat in a little olive oil a little salted pork, or 3 slices of bacon. Add 1 medium-sized onion cut in very thin slices, add 8 or 9 cups water, salt and pepper, a handful of pounded almonds, a clove of crushed garlic and 4 large potatoes cut as for frying—lengthwise—about 10 slices to each potato. When the potatoes are cooked add the juice of 1/2 lemon and 1 wine-glass Malaga or any other sweet white wine, and 2 pinches of saffron. Finally drop 1 egg per person into the soup and let poach and serve.

Found and eaten with delight in the island of Ibiza.

Algonquin Hotel
New York

These two soups supplied by the *chef* through the kindness of the so amiable *maître d'hôtel* Georges at Gertrude Stein's request.

CREAM OF FRESH MUSHROOM SOUP

An equal amount of onions, leeks and celery chopped fine. Also fresh mushrooms. Mix together in a pot with butter and simmer partly. Then add 1 or 2 tablespoons flour. Stir while heating through. Add stock and cool for 1 hour. Strain and add some cream.

ALGONQUIN SPECIAL SOUP
(serve cold)

Put some butter in a pan with some Indian curry powder. Stir until curry and butter are mixed well. Then add some chicken broth and let boil for 10 minutes. Finally mix the yolk of 1 egg with plain cream for each cup of soup.

ENTRÉES

Sir Francis Rose, Bart.
Paris

CHINESE EGGS

1. Boil eggs for 5 minutes.
2. Remove shells carefully and cook in sherry for 5 minutes.
3. Put eggs in covered casserole with butter and 3 large spoonfuls soya sauce, and cook until dark brown.

From the Same Source

LEMON SALAD

Boil lemons until soft in a lot of very salty water. Drain and let cool and cut them. Mix them with cooked cut-up artichoke stems and hearts, salted almonds, 1 coffee-spoon honey and 2 large spoons oil. Sprinkle with lemon juice.

Dr. Fernanda Pivano-Sottsass
Milan

GNOCCHI ALLA ROMANA

3/4 lb. semolina.	5 ozs. Parmesan cheese.
3 ozs. butter.	1 glass milk.

Cook the semolina in water and milk for 40 minutes. Spread it on a wet marble slab stretching it well with a wet wooden spoon until it is about 2/3 inch thick. When it is quite cold divide it into small rounds. Grease a baking tin and put the gnocchi on it, spreading them with butter and grated cheese. Continue to put layers of them, covering each with butter and cheese. Put the baking tin in already warmed oven and leave for 30 minutes.

Virgil Thomson
New York

SHAD-ROE MOUSSE

Remove the flesh from 2 lbs. fresh halibut and put the bones, odd bits, and skin into an enamel saucepan with a small lump of butter, onion chopped fine, a little bouquet of parsley, a bay leaf, a tiny pinch of thyme, a few peppercorns, 3 cups water and 1 cup white wine. Put on the fire to simmer. This liquid is to use later in the making of a sauce. Put the raw flesh through the meat grinder, add the unbeaten whites of 6 eggs, and mash the whole thing through a very fine sieve. Add some salt and pepper and stir until it thickens a bit, then gradually add 1 pint cream. In the meantime, cook for about 20 minutes 2 or 3 pair shad roes slowly in a frying pan with plenty of butter. Carefully remove the skin and veins. Mash the roe lightly and add to the fish-paste. Mix well. Add another cup of cream and then season to taste.

Butter two medium-sized fish-shaped moulds and decorate the bottom with strips of red pimiento. Fill the moulds with the *mousse*, packing it well down into all the crevices. Set the mould into the refrigerator and keep there until 40 minutes before you will be ready to serve. At that time place the moulds in a pan of hot water, cover with a piece of buttered white paper, and set in a moderate oven (about 400°) to cook.

In the meantime, make the foundation for your sauce by putting 1/2 cup butter into an enamel pan to melt. Stir in 3/4 cup flour and cook for a minute or two without browning. Add gradually the strained fish stock of which there should be about 3 cups. Continue to cook in a double boiler until ready to serve. Then, add the yolks of 2 eggs beaten up with 3/4 cup cream, stirring continuously and making sure it doesn't cook any more. At the last minute, add a little paprika and stir in 1 cup good sherry and the juice of 1 lemon. Pour around the *mousses*, which have first been carefully emptied from the moulds on to warm platters. Garnish and serve at once.

June Platt recipe

From the Same Source

GNOCCHI ALLA PIEMONTESE

2 lbs. white potatoes (of the kind which fall to bits).
5 ozs. fine flour.

1/2 lb. fontina (soft, melting cheese).

1 egg.

3 1/2 ozs. butter.

Boil the potatoes for about an hour. Peel them and sieve them or mash them. Put them on the table and knead them with flour, adding the egg. Divide the mixture in small pieces, the size of walnuts. Curl them on the fork.

Meanwhile you'll have prepared the fontina, sliced finely and put in milk for about an hour to make it very soft. Plunge the gnocchi in a saucepan of boiling water and leave them to boil for about 3 minutes; they will come to the top by themselves. Get the dish ready with a layer of the finely sliced, milked fontina and butter and put the strained gnocchi in layers, covering each with fontina and butter, and mixing each layer before doing the next one.

From the Same Source

PORK "ALLA PIZZAIOLA" OF CALABRIA

Enough for 5 people.

1 lb. sliced pork.
1 tin (2 ozs.) peeled tomatoes.
1 oz. salted anchovies.
1 oz. capers.
2 ozs. parsley.
A pinch of salt.

Brown the pork in 3 tablespoons butter and take it off as soon as done. Finely mince the capers, anchovies and parsley and put them into the butter. Add the tomatoes and about 1 wine-glass water. When the mixture begins to boil, add the pork which was put aside and cooked for an hour on a slow fire.

MADAME GASTON CHABOUX
Belley

TRUFFLE TURNOVERS

Prepare a short or puff paste. Roll it out to 1/6 inch thickness. Brush and wash the truffles without peeling them, cut them in dice of 1/6 inch thickness, place them on a slice of fat side of pork which has been heated in a frying pan, cover the truffles with a very thin slice of Swiss cheese, fold them in the dough in the shape of turnovers. Bake in the oven or fry in deep fat.

FROM THE SAME SOURCE

CHICKEN-LIVER CUSTARD

Pound in a mortar 2 or 3 chicken livers, add 4 or 5 eggs, 2 cups milk. Mix thoroughly, salt and pepper. Strain through a fine sieve. Pour into an oiled mould. Cook in the oven, the mould standing in a pan of hot water. Serve with a tomato or crawfish sauce.

GEORGES MARATIER
Paris

LIVER CUSTARD

1. Prepare a fairly thick *Béchamel* sauce.
2. Two chicken livers, or a rabbit's.
3. Chop fine: parsley, a clove of garlic.
4. Mix these together.
5. Add 3 yolks of eggs. Beat the whites, and pour all this into a well-buttered mould.
6. Put in a *bain-marie* in a medium oven for 1/2 hour.
7. Prepare a hot tomato sauce with butter and flour, black olives and chopped mushrooms.
8. Remove custard from the mould and cover it with the sauce. Serve hot with a cold Alsatian wine.

FROM THE SAME SOURCE

VEAL KIDNEYS

1 veal kidney.
2 tablespoons butter.
1 dessertspoon cornflour.
1 tablespoon mustard.
1/4 cup gin.
Salt.
Pepper.

Clean and cut the kidney in dice. Put butter the size of a walnut in a *cocotte*, and lightly brown the kidney. Light the gin in the *cocotte*, having previously removed the blood. Cover the *cocotte* and remove from fire.

Make a sauce by melting the rest of the butter, add the cornflour which has been mixed with the blood, then add the mustard, salt and pepper. Pour this mixture into the *cocotte* and finish cooking without permitting it to boil.

MARY OLIVER
London

FONDUE DE BÂLE

Heat 1 lb. Gruyère cheese and 1 dessertspoon butter over chafing dish. Stirring carefully, add 1/2 bottle Riesling wine. Keep stirring, add 1 liqueur glass kirsch. Continue to stir and add 1 teaspoon powdered mustard. Serve with toast or bread cut in small squares. Place bread on fork and dip bread into *fondue*. Keep *fondue* warm but not so hot as to scorch. Riesling wine well chilled should be drunk with this. *Fondue* should not be served on plates. Each person should dip his bread into the communal bowl.

FISH

Mrs. Gilbert Whipple Chapman
New York

SHAD ROE

1/2 cup lemon juice.	1/2 bottle walnut catsup.
3/4 cup butter.	1/4 bottle (small size) Worcestershire sauce.
Salt and pepper.	2 pairs shad roe.

Parboil the shad roes and cut them in half lengthwise. Lay aside to keep warm. Melt the butter in a frying pan, and pour in the lemon juice and cook for nearly 20 minutes until the acid liquid has boiled off the lemon juice and there remains a light-brown syrup. Pour in 1/4 bottle Worcestershire sauce (small size) and 1/2 bottle walnut catsup and cook for 2 or 3 minutes. Lay in the shad roes face up and cook for 2 or 3 minutes, basting at the same time. Place the shad roes on a platter and pour in a little more walnut catsup and cook sauce for a couple of minutes more and pour over shad roes and serve.

Lady Rose
Nice and London

BARRIDA
(*a Corsican fish dish*)

For 8 persons: Make a mayonnaise with 1 quart olive oil, 12 yolks of eggs, 12 cloves of garlic. Boil 1 turbot or fresh cod, 1 sea eel or sea perch, and any other fish from Southern waters. Stir the water in which the fish has boiled into the mayonnaise so as to make a sauce the consistency of cream. Serve the fish and sauce with boiled new potatoes, carrots, string beans and hearts of artichokes. The sauce looks like yellow cream and is heaven.

The meal commences with a soup made of the sauce further

watered down with the water in which the fish has been boiled, with pieces of bread floating in it—pale yellow, nearly clear and delicious. With this is served a sparkling white wine, very dry. Really a lovely meal.

MARY OLIVER
London

PILAW STELLA MARIS DE PORTO FINO

Cut up one small octopus, remove bone from interior. Dip particles in honey, roll in paprika, then plunge in batter mixed with garlic. Boil in olive oil. Serve with rice; with a sauce made with tomatoes over it, white wine, green peppers and finely diced mushrooms.

PRINCESS D. DE ROHAN
London

SCAMPI PINO ORIALLI

Place scampi, Dublin Bay prawns, langoustine or giant shrimps on a mound of rice which has been washed in fourteen waters.

Cook scampi in cold water . . . Bring to the boil . . . Add cold water.

Cook rice in cold water . . . Bring to the boil . . . Add cold water . . . Place in oven to dry. To serve, add shell-fish.

Serve with a sauce made of tomatoes, 5 big onions, 12 cloves and saffron to taste. On top of which add whipped cream.

LADY ROSE
Nice and London

MUSSELS WITH RICE
(a Corsican dish)

For 3 people: Clean 2 lbs. mussels and put them in covered casserole on fire until they open. Remove from shells. Pass juice from the mussels through a sieve. Cook rice (2 soupspoons per person) in casserole in olive oil until lightly golden. Cover rice with juice of mussels mixed with water, 4 soupspoons of liquid to 1 soupspoon rice. When water boils, lower heat and simmer for 1/4 hour. The liquid should by then have evaporated. Mix rice with mussels and serve.

MAURICE GROSSER
New York

SHRIMPS IN BUTTER

Fry a chopped onion in sweet butter, add peeled raw shrimps and the juice of 1/2 lemon (or more, according to taste). Cook for a few minutes until the shrimps have turned pink—no longer—sprinkle with chopped parsley, and serve on rice.

If the shrimps are fresh, unfrozen and without the awful preservative they are sometimes treated with they are unbelievably good.

FROM THE SAME SOURCE

BAKED FISH WITH STUFFING

A large fish, red snapper or red bass, stuffed with oysters, chopped onion and celery, the latter fried previously lightly in butter. The fish is baked and basted with grapefruit juice. In the dressing use parsley and a little bay leaf—no thyme or sage.

MARY OLIVER
London

FILET DE SOLE À LA RITZ

Poach filleted sole in frying pan half full of water. Serve cold with sauce of whipped cream which has been mixed with grated horseradish.

MEAT AND GAME

HAROLD KNAPIK
Paris

This is the goulash that I mentioned. It is not bad but its origin on the Hungarian plain is reflected a little insistently.

SZEKELY GULYAS

1 large onion, chopped.
3 tablespoons butter.
1 green pepper, chopped.
5 medium tomatoes, peeled.
1/2 lb. veal, cut in strips.
1/2 lb. tender beef, cut in strips.
1/2 lb. pork, cut in strips.
(Cut excess fat from the meat.)
2 1/2 teaspoons salt.
6 peppercorns.
2 bay leaves.
1/2 teaspoon capers.
1 tablespoon paprika.
1/2 cup stock.
1 1/2 lbs. sauerkraut.
1 1/2 cups sour cream.
1 teaspoon caraway seeds.

Fry the onion until brown and then add the tomatoes and green pepper. Cook for about 15 minutes, very slowly, and then add the meat and the seasonings. Add the stock and simmer, covered, for 1/2 hour. Then add the kraut and cook for 1 hour longer. Before serving, add the sour cream.

FROM THE SAME SOURCE

VEAL-CHOPS PAPRIKA

To my knowledge this is my own recipe, but its resemblance to goulash is evident.

The ingredients for 4 people are:

4 thick, first-quality veal chops.

1/4 lb. very thinly sliced Hungarian bacon which is fat back, lightly smoked, with paprika pressed into the fat. Anything comparable will do in its place.

1 medium-sized onion, finely chopped.

1 1/2 teaspoons paprika. Real paprika, and not red dust, is required.

1 1/2 cups chicken stock.

1 cup sour cream.

1/2 teaspoon salt.

Lightly brown the bacon, push the bacon aside and brown the chops well. Turn down the fire and brown the onions gently with the paprika. Add the heated stock and salt and cook, covered, over a low fire for about 20 minutes or until done. Just before serving add the sour cream. A few *croûtons* fried in butter and a little chopped parsley provide a reasonable garnish for this dish.

From the Same Source

This is a Turkish dish, *considerably modified.* Combining them with the rice and the sauce is, for better or for worse, my own. In Turkey, the *Kebabs* are eaten alone as an appetiser.

KEBABS WITH RICE

This is a modified Turkish *Kebab.* It is usually made or formed on a number of small flat spits which unfortunately are very difficult to find. However, if the meat does not contain too much fat and the onions are chopped really fine, one can roll long, slender, pencil-like *Kebabs* that hold together very well. The ingredients for 4 people are:

2 lbs. shoulder of lamb or mutton, with a third of the fat rejected, and ground very fine.

2 medium-sized onions chopped as finely as possible.
1 1/2 tablespoons powdered cumin.
1/2 teaspoon salt.
1 strong pinch of cayenne pepper.
4 tablespoons mixed green herbs.

Mix vigorously all the above ingredients for at least 10 minutes, then roll the *Kebabs*. Place them on a grill and cook gently but well. If cooked too much they will dry. Turn over once while cooking.

They can be eaten with rice prepared in any manner, and are quite good with a sauce made as follows:

Fry gently in 1/4 lb. butter a finely chopped onion, a finely chopped green pepper and 1 tablespoon saffron. Cook for about 15 minutes and mix with some of the drippings from the *Kebabs*. Place the rice on a platter with the sauce thrown over it and the *Kebabs* arranged around the rice. Place quarters of lemon between the *Kebabs*.

Mrs. Noel Murphy
Orgival

RABBIT WITH DUMPLING

Cut your Belgian hare!! in pieces. Roll in flour and brown in an iron pot in which you have slightly cooked 4 or 5 slices of bacon, adding bacon fat or lard. Put in a bouquet of herbs (I prefer rosemary, laurel and the greens of fresh garlic) and an onion stuck with cloves. When golden brown add thinly cut carrots, dust again generously with flour and add red wine and water. Cover and let simmer slowly. When cooked the meat should have the consistency of chicken, and not slimy restaurant rabbit. Add 1/2 teaspoon Worcestershire sauce, more water or *bouillon* if necessary. There should be much sauce. Add 1/2 cup cream before serving.

Dumplings

Make a dough with 2 cups flour and 1 egg, 1/2 teaspoon salt and water. It should be like bread dough. Cut up *very stale* bread in squares to make 1 cupful. Mix thoroughly with dough and make oval balls. Boil in salted water for 1/2 hour. When drained cut immediately into slices. The dumplings should not be put into the rabbit sauce, but the sauce put over the dumplings at table.

Redvers and Louise Taylor
Bishops Lydeard

JUGGED HARE

Skin and clean, chop in small pieces. To 1 hare and its blood add 1/4 of a bottle of red wine, 1/2 gill vinegar, 1 large onion cut into six pieces, 2 carrots, a bay leaf, 12 peppercorns, pepper and salt well (at least a heaped teaspoon of the latter). Mix in a bowl. Press well down. Cover over and let stand for 24 hours, to pickle.

Put into a colander and strain dry. Fry in a frying pan (till brown) everything that was left in the colander. Then place in a casserole, sprinkle with 1 large tablespoon flour. Add 1/2 pint water and all the liquid that had strained through the colander and 1 tablespoon currant jelly.

Cook slowly from 40 minutes to 3 hours, depending on the age of the hare.

Fania Marinoff
New York

LAMB CURRY FOR SIX

3 lbs. lean lamb, in small pieces.	2 glasses red wine.
7 small apples.	2 tablespoons cooking oil.
4 onions.	2 tablespoons flour.
2 cups stewed tomatoes.	2 tablespoons curry powder.
2 cups rice.	1 cup raisins.

Mix uncooked lamb with curry powder. Season with salt and pepper. Brown the apples, onions and tomatoes in the oil and wine, add the uncooked lamb and cook slowly for 1 hour. Drain off juice, thicken with flour and return to pot. Cook rice separately and mix with raisins. Serve with chutney.

MARY OLIVER
London

ROAST PORK NORMANDY

Roast pork in 1/2 bottle cider. Baste constantly and keep adding more cider. Serve with fried apples that have been well sugared and dusted with powdered cinnamon. Fry apples in bacon fat.

FROM THE SAME SOURCE

ROAST BEEF FOR A RAINY DAY

First lard the roast with garlic, then soak for 2 hours in the contents of a bottle of sweet red wine and bay leaves. Place in basting pan with the wine poured over the meat. Cook in slow oven and baste every 15 minutes. Serve with the wine gravy.

VEGETARIAN

MERCEDES DA ACOSTA
Paris

STUFFED ARTICHOKES STRAVINSKY

Before cooking put a little garlic and lemon juice inside artichokes. Cook artichokes until tender, then take off leaves and put hearts in a baking dish. Prepare a fresh mushroom sauce as follows:

Separate the tops of the mushrooms from the stems. Mince the stems and dry them in the oven. Slice the tops of the mushrooms and sizzle in butter rapidly. Powder with flour, using flour sifter. Mix well and add 2 tablespoons sour cream. Season to taste. Stuff the artichokes with this mixture and powder it with dried, cut mushroom stems. Bake for 5 minutes.

SPANISH RICE

Cover bottom of pan with melted butter and oil, then cut purple onions into very very thin slices and place in pan. Put 1 handful of rice per person, and 1 for the pan, in hot water, then strain, mix well with oil and chopped pimiento. Put in pan with onions and cover, cooking in a very slow oven until done. If desired, tomato can be added.

CUBAN RICE

2 cups M.J.B. rice.
4 cups water.
6 tablespoons olive oil.
1 clove of garlic, chopped fine.

Mix well. Place in a *very slow* oven in covered kettle. Do not stir while baking. Cook for 35 minutes.

VEGETABLE ROAST LOAF

(recommended to confirmed meat-eaters who would like to imagine they were eating meat)

1 medium-sized egg plant.
1 cup ground celery.
1 cup cottage cheese.
1 tomato.
1 tablespoon wheat germ.
4 tablespoons butter.

1 teaspoon ground pimiento.
1/2 teaspoon Savita.
1 egg.
1/2 cup soya-bean bread-crumbs.
1 ground medium-sized onion.

Peel the egg plant and put it and vegetables through a food grinder. Melt 1 tablespoon butter in a frying pan, put all the vegetables in it and simmer until the vegetable juice is cooked in. Mix in Savita, remove to a mixing bowl and let cool. Add beaten egg, breadcrumbs, wheat germ, cottage cheese and 2 tablespoons butter. Butter baking dish and dust with breadcrumbs; pour mixture in it, and cover with breadcrumbs and melted butter. Bake in a hot oven for 25 minutes.

CHICKEN AND A BIRD

Pierre Balmain
Paris

"VENT VERT" CHICKEN

Choose young chickens of about 1 lb. each (2 chickens for 3 people). Carve them uncooked as if they were cooked (legs, wings and breast—put the carcass and the giblets aside). Make a strong *bouillon* seasoned with thyme, laurel, cloves, onions, celery salt and Indo-Chinese pepper.

Half an hour before serving, *sauté* in butter over low heat for about 1/2 hour the pieces of chicken, with salt and pepper. To serve place on a silver dish. Pour the prepared *bouillon* with a dash of brandy into the casserole. Stir to incorporate the glaze into the sauce. Then add a large handful of chopped fresh tarragon and let it come to the boil. Cover the pieces of chicken, placed on the silver dish, with this tarragon sauce. Fresh cream may be added according to taste. (Personally I prefer the chicken without cream.)

Chicken to be served without vegetables, but to be accompanied by

"Vent Vert" Salad

Cut in small right angles, not omitting the stalks, the hearts of young romaine salads. Add celery, endives, sweet green peppers and asparagus tips, a quarter of the volume of the romaine salad for each of these last ingredients. Add to this some leaves of corn salad from which the stalks have been removed to make *little perfect* ovals and about 1 tablespoon per person of Beaufort cheese, savoyard cheese of the Gruyère type cut into thin matches.

Prepare a salad sauce with salt, Indo-China pepper, white-wine vinegar and walnut oil (1 tablespoon vinegar to 2 tablespoons walnut oil). Carefully peel 2 fresh walnuts per person, cut them into very small pieces and sprinkle on the salad, which is to be served in a green unbaked earthenware salad bowl.

Redvers Taylor

Bishops Lydeard and London

CIRCASSIAN CHICKEN

Fry 1 or 2 onions in 1/2 tablespoon butter until they begin to colour. Separately, boil a fowl and add the onions. Take 10 ozs. shelled walnuts and pass them through a mincing machine. Then mix in well 1 teaspoon red pepper.

Take the crumbs of half a loaf of bread and soak them in water in which fowl was boiled. Work them in with the walnuts with a wooden spoon, adding a little salt. If the mixture is too dry, gradually add the water in which the fowl was cooked until a thick cream-like consistency is obtained. Bone the fowl.

Arrange the filleted fowl in a dish and pour over the sauce. Decorate and serve cold with salad.

Madame M. G. Debar
La Régie
Soye-en-Septaine

MESSY CHICKEN À LA BERRICHONNE

Brown a chicken cut in pieces in butter and diced fat back of pork—half the pork fat should be fresh, the other half smoked—chopped garlic, onions, shallots, parsley, chervil, tarragon, and a bouquet. Moisten with 1/4 cup brandy, 1 1/2 cups red wine of Bordeaux or of Burgundy, or another very good red wine, salt and pepper. At the end of the cooking, add 1/2 lb. mushrooms heated in butter. Thicken the sauce slightly with flour, add the blood of the chicken in which a few drops of vinegar was mixed when the chicken was killed. Add slowly to the sauce 1 cup cream in which the yolks of 2 eggs have been stirred. Serve surrounded by little puff-paste crescents.

The Late Lord Berners

ROAST CHICKEN IN CREAM

Brown 2 or 3 onions in butter, add chicken. When cooked remove chicken and keep hot. Add some cream to a little of the butter in which the chicken has cooked. Add salt, pepper, a little lemon juice, a little sherry or Madeira. Let the sauce reduce until it begins to thicken. Then carve the chicken, and pass the sauce through a very fine strainer over the chicken and serve.

Néjad
Paris

BOILED CHICKEN

Cut 2 chickens into very small pieces. Put in a saucepan and cover with 5 cups and 3 tablespoons milk, add 1 3/4 cups sugar and 1 tablespoon rose water. Twenty minutes before the chickens are tender, add 2/3 cup rice. When the rice is cooked, serve.

PRINCESS D. DE ROHAN
London

CHOP SUEY
(*from chicken left-overs*)

(Recipe from Chong Ping Nam, one-time *chef* to the Chinese Ambassador.)

Dice chicken. Heat in soya sauce and butter. Add 1/2 pint peeled prawns or shrimps. Serve with bowl of rice and a cup of China tea, hearts of lettuce salad, lemon and oil dressing.

Diced roast pork can be used instead of chicken.

MARY OLIVER
London

LARKS À LA CONCHITA HERNANDEZ
(*a gypsy singer of Madrid*)

Place 2 dozen plucked larks in an oven with 6 rashers of Parma smoked ham or bacon and serve on platter in a bed of watercress. Surround by raw Spanish onions, raw tomatoes and red and yellow pimientos preserved in oil.

JOSEPH DELTEIL
Grabels

CHICKEN WITH RICE

Brown your chicken, then add an onion with 2 cloves stuck in it, 2 carrots, a bouquet, 2 cloves of garlic, 1 glass white wine, 1 quart water, salt and pepper. Let it boil. If she is not too old (say forty years) the hen should cook in an hour.

Apart, melt 1 tablespoon butter in a casserole. Add 1 finely chopped onion, stir over the fire, put into it 1 cup rice, give it a moment to heat and then add half the juice in which the chicken has cooked and which has been strained. Add salt and pepper, nutmeg and a good pinch of saffron. Let it cook covered very gently for 25 minutes.

Apart, put a piece of butter the size of a walnut in a saucepan. Add 1 tablespoon flour, stir a moment over the fire without allowing it to brown. Add the rest of the *bouillon* of the chicken and bring it to a boil. Add a thickening of the yolks of 2 eggs (from the same chicken), a little lemon juice, and serve your chicken with rice.

And good appetite to you.

SAUCE

Dr. Fernanda Pivano-Sottsass
Milan

PESTO ALLA GENOVESE

(*sauce which can be served to flavour thick vegetable soup or pasta asciutta*)

Enough for 4 people.

3 ozs. good fresh basil.	2 ozs. Parmesan cheese.
1 oz. parsley.	1 oz. pine-nuts.
2 ozs. Pecorina cheese (strong *goat* Parmesan).	2 ozs. butter.
	3 1/2 ozs. oil.

And, if liked, a clove of garlic.

Clean the leaves of the basil and parsley and leave them in water to keep them fresh. Grate the two cheeses together, mince the leaves of basil and parsley and the pine-nuts together (with the garlic, if you like it), add the cheese slowly so as not to blacken the mixture. The mixture has to be very light. Put it all in a bowl and add the oil, continuing to mix it. Just before using add the butter and a ladle or two of the boiling liquid which is to be flavoured (for instance, water of the pasta asciutta or broth of the soup).

VEGETABLES

Miss Natalie Clifford Barney
Paris

STUFFED EGG PLANT WITH SUGAR

2 egg plants.	2 dessertspoons sugar.
1/4 lb. dried breadcrumbs.	1 large pinch of salt.

Divide the egg plants in half lengthwise. Remove the pulp, chop, add breadcrumbs, sugar and pepper. Stuff the four halves very abundantly. On each one place a piece of butter the size of a large walnut and 1 tablespoon of water and cook in a moderate oven for 1/2 hour.

The Late Lord Berners

STEAMED SPRING VEGETABLE PIE

Line a pie dish with puff paste, reserving the quantity necessary for a lattice. Bake the crust. Steam all the spring vegetables available—peas, onions, carrots, string beans, asparagus tips and so on. As soon as the crust is baked, place on it the vegetables in bouquets of each kind, and sprinkle with a small quantity of butter. Bake until lattice is golden brown. Serve at once.

Miss Katherine Dudley
Paris

BROWNED-IN-THE-OVEN WHITE BEANS

Boil in water an onion with a clove stuck in it, 1 quart good white beans, fresh or dried; if the latter, they must be soaked for 24 hours in fresh water. They must cook very gently so that they will remain whole.

Put in a casserole a spoonful of good *Béchamel* sauce and a pint of cream, a spoonful of grated Swiss cheese, and a seasoning of salt and pepper. The sauce should be creamy but not thick. Drain the beans thoroughly. Place them on a deep fireproof dish. Cover them with the sauce in which they are cooked. Cover with grated cheese and put the dish on the grill of the oven to brown.

Sir Francis Rose, Bart
Paris

STUFFED ITALIAN SQUASH (ZUCCHINI)
(*a Chinese dish*)

1. Blanch a very large Italian squash. Cut off the top and scrape out interior.
2. Fry half-cooked rice in butter with chopped meat for a few minutes and mix with chopped basil and onions.
3. Stuff the marrow and tie it up well, covering it with bacon and sprigs of parsley. Bake until bacon and parsley are crisp, basting often.
4. Shrimps or prawns are added in the Chinese style:

Shell the shrimps or prawns and put in earthenware dish. Cover with 1 large spoon honey and 1 large spoon sherry (or Suchow spice wine). Then add 2 spoons soya sauce. Let the dish stand for several hours, then put in oven for 25 minutes. Remove, add a covering of chopped spring onions and put back in oven for 5 minutes.

MARY OLIVER
London

STUFFED PEPPERS HAMMAMET

Boil barley in salted water until tender—it should absorb all the water. Mix with chopped onions and parsley. Fill green peppers with this mixture, cover with olive oil, and put in oven. Serve with sauce made of lemon juice and paprika.

FROM THE SAME SOURCE

MASHED POTATOES LUXEMBOURGEOISE

Mash potatoes in butter, and red wine instead of milk.

MADAME GASTON CHABOUX
Belley

STUFFED SWISS CHARDS

Parboil the leaves of some Swiss chards. Prepare a forcemeat of left-over roast—preferably mutton or chicken. Prepare a brown sauce with 2 tablespoons butter and 2 tablespoons flour, the gravy of the meat and cream added in small quantities at a time, salt and pepper, spices. In this sauce place the meat (chopped), and on the leaves of the Swiss chards place a tablespoon of this mixture. Roll and fold the edges. Put in a dish and cook in the oven.

MADAME BERTHE CLEYSERQUE
Paris

SAUTÉ OF MIXED VEGETABLES

(a Roumanian dish)

It will require:

1/2 lb. veal. | 2 Italian squash.
2 tomatoes. | 2 mushrooms.
2 onions. | 3 ozs. tomato *purée.*

All of medium size.

The vegetables must remain whole. Cut out the insides, add to the forcemeat salt, shallots and parsley, all finely chopped, including the veal. Fill each vegetable. Put in a casserole a piece of butter a little larger than a walnut, brown the vegetables without turning them. Add the 3 ozs. of tomato *purée.*

Allow to simmer very gently for 1 hour. Just before serving pour 1/2 cup cream over the vegetables and pour the sauce over it. Serve very hot.

SALADS AND SALAD DRESSING

Carl Van Vechten
New York

GARLIC ICE CREAM

(*a dressing for salad*)

4 small tomatoes, chopped to pulp.
1 tablespoon Worcestershire sauce.
1 teaspoon tabasco.
1/2 teaspoon salt.
1 teaspoon onion juice.
1 cup mayonnaise.
2 spoons Cowboy's Delight (may be procured from Old Smoky Sales Co., 124 West 4th Street, Los Angeles, California).

Beat till ingredients are well mixed. Freeze in icebox. DO NOT STIR WHILE FREEZING. Serve in avocados (cut in half).

Mrs. Noel Murphy
Orgival

SAFFRON RICE

Boil 2 handfuls of rice per person with a heaped soupspoon of powdered saffron and 2 cloves of garlic which you remove. (The rice should be yellow when cooked.) Do *not* overcook the rice. Take the stones out of 5 ozs. black olives, add 1/4 lb. shelled shrimps, about 6 fresh pink *raw* mushrooms cut in thin slices.

Make a generous French dressing with much olive oil and little vinegar and finely chopped chervil. Mix with rice and garnish with sweet peppers.

Mary Oliver
London

14TH OF JULY SALAD

To a pint of mayonnaise add capers and chopped dill pickles. Mix well with 1 lb. boned white fish. Serve with a salad of nasturtium leaves and cucumbers with a dressing of olive oil and garlic mixed with tarragon vinegar. Garnish dish with nasturtium leaves and orange and red nasturtiums.

With this should be served chilled chives, or a cider cup with raspberries and cucumber rinds.

Princess D. de Rohan
London

SALADE APHRODITE

Apples, quickly peeled and finely chopped, celery chopped fine, yoghourt, black pepper, salt.

The beauty of this salad depends entirely on how quickly the apples and celery are stirred *into* the bowl of yoghourt. This prevents their becoming brown. To be served on the crispest lettuce leaves.

This is inspired by the famous "Bicht's Mousslé" of the Bicht Sanatorium at Zürich. Ideal for poets with delicate digestions.

Mrs. Carlton Lake
Paris

ASPIC SALAD

1 can Campbell's Condensed Tomato Soup.
1 small package Philadelphia Cream Cheese.
2 tablespoons unflavoured gelatine dissolved in 1/3 cup cold water.
1/2 cup Miracle Whip Salad Dressing.
Chopped vegetables: 1/2 onion, 1 stalk of celery, 1 green pepper.

Empty tomato soup into saucepan, and heat over low flame, stirring until soup is almost hot. Add the cream cheese, and keep stirring until it is dissolved. Remove from fire, and add gelatine which has been dissolved in water. When mixture cools, fold in the salad dressing. Chop the vegetables—onion, celery, green pepper—and when mixture begins to set, fold in. Rinse out ring mould in cold water, but do not wipe dry. Pour mixture into mould, and chill in refrigerator for several hours. Serve on bed of lettuce leaves.

To serve, loosen salad along edges with sharp knife, invert, and tap mould. If salad does not come out of mould easily, wipe outside surface of mould with cloth wrung out in hot water.

BREAD AND CAKES

Mrs. Redvers Taylor
Bishops Lydeard and London

CRULLERS

Beat the yolk of 1 egg until stiff. Stir into it 2 heaped tablespoons sugar and 2 tablespoons melted butter. When very light beat the white of the egg very stiff and blend it with the mixture. Add nutmeg and salt. Mix to a stiff dough with flour, using enough to enable you to roll it out, 1/3 inch thick. Cut in squares, make three or four long incisions in each square. Cook in hot fat and sprinkle with sugar. In a stone jar they keep crisp for ages.

Fania Marinoff
New York

PECAN NUT CAKES

3 cups pecan nuts, chopped as finely as possible.
6 eggs.
1 1/2 cups sugar.
1 tablespoon flour.
1/2 teaspoon salt.
1 teaspoon baking powder.
1 teaspoon vanilla.
1/4 cup chopped seedless raisins.
Boiled white icing sugar.

Beat yolks until light, add sugar and beat again. Mix the finely chopped nuts, flour, baking powder and salt, and add to the yolks. Beat well and stir in the stiffly beaten whites. Bake in two layers in moderate oven. When cold, assemble and ice with boiled white icing sugar, sprinkling a few chopped raisins on top.

Néjad
Paris

COOKIES

Boil in a saucepan 1/2 cup and 2 tablespoons butter, 4 cups water and a good pinch of salt. Moisten with some of the water 5 cups rice flour. Gradually add it to the boiling water, stirring with a wooden spoon. Continue to stir until the mixture becomes stiff. Remove from heat and very slowly add 10 eggs one at a time. Incorporate each one thoroughly before adding another. Roll them into little sausages and place on buttered baking sheet in a moderate oven. When cold paint with a water icing flavoured with rose water.

Redvers Taylor
Bishops Lydeard and London

BANBURY CAKES

Prepare puff pastry; 1/2 lb. flour will make enough for 20 cakes. Filling for cakes: 1/2 teaspoon ground ginger, 1/2 teaspoon lemon rind and juice, 1 oz. candied peel chopped, 1 oz. cake crumbs (or breadcrumbs with sugar added), 2 ozs. sugar, 2 ozs. sultanas, 2 ozs. currants, 1/4 teaspoon ground nutmeg, 1/4 teaspoon cinnamon, 1 egg, 2 ozs. butter. Warm bowl and then work butter, sugar and the spices into a creamy consistency. Then beat in egg and later the crumbs. Then add lemon juice, currants, etc., and lemon rind. Stir.

Cut pastry into rounds. Put some of the filling on to a round. Fold over to make a semi-circle. Egg-wash the joint and squeeze tight the ends. Place it joint downwards and tap it out to an oval shape. Make a few cuts in top. Bake for 20 minutes at 500°.

Dr. Fernanda Pivano-Sottsass
Milan

PIZZA ALLA NAPOLITANA

2 cups flour.
1/3 oz. yeast (either of beer or bread).
5 ozs. Mozzarella (very soft, melting cheese).
1 two-oz. tin peeled tomatoes.
3 salted anchovies.

Knead the flour and the yeast together on the table with 1/2 glass milk, adding water if necessary, with a pinch of salt. Let it rise for 45 minutes, wrapping it in a woollen cloth. Stretch it into a round shape, and put it in a flat greased baking tin. Sprinkle on the mixture 3 tablespoons oil, pressing it with fingers. Spread on top first half of the thinly sliced Mozzarella, second the tomatoes halved, third the cut-up anchovies, and finally a pinch of origan and the rest of the Mozzarella.

Put it in the oven already warmed and cook for about 30 minutes, according to the heat of the oven.

MADAME GASTON CHABOUX
Belley

A GARNISHED SALTED BREAD OF THE BUGEY

On an ordinary bread dough place the following mixture: 1 or 2 minced onions, 2 or 3 tablespoons chopped walnuts, 1 cup walnut oil. Send it to the baker's to be baked in his oven after the bread has been removed.

CRÊPES AND PANCAKES

MRS. GILBERT WHIPPLE CHAPMAN
New York

CRÊPES NORMANDES

Melt a very thin layer of butter in a small frying pan over a hot fire. Spread 4 or 5 very thin slices of green cooking apples in the butter. Cook for 2 or 3 minutes. Pour over these a slight coating of thin pancake batter. Cook this a minute or two longer. Pour on this 2 heaped tablespoons sugar, and cook for 2 minutes more. Add

1 tablespoon butter, and turn the pancake over, and cook a little longer over a low flame, adding a little more butter. Take out and serve when fairly crisp. This makes one portion.

From the Same Source

SALZBURGER NOCKERL

Made the easy way. This recipe is for people who can not toss a *soufflé* omelette in the air to turn it over in the pan.

Mix 1 1/2 tablespoons flour and 4 tablespoons granulated sugar, and 1 pinch of salt. Add these to the well-beaten yolks of 6 eggs. Blend well, and then fold into the well-beaten whites of 6 eggs. Melt 1/4 lb. butter in a large, deep, iron frying pan. Pour the mixture into this. Cook over a slow flame for 3 to 4 minutes. Then place under the broiler and cook slowly for 3 to 4 minutes longer. Put in a slow oven for 2 to 3 minutes longer. While you are cooking this, melt 1/2 lb. butter until it is brown. Turn your *soufflé* over on a hot dish, and immediately pour the brown butter over it, and sprinkle well with icing sugar. Serve immediately.

Carl Van Vechten
New York

VIENNESE CHEESE PANCAKES

2 yolks of eggs.
1/2 teaspoon salt.
2 teaspoons sugar.
2 cups milk.
1/2 cup, or more, flour.

Beat the yolks of the eggs and pour all together in large bowl. Make THIN pancakes and fill them with: pot cheese, raisins, yellow of one egg, vanilla, sugar. Bake for 10 minutes in rich butter.

Now, I hope, you have had enough recipes from US.

DESSERTS

Sir Francis Rose, bart
Paris

QUEEN ELIZABETH I APPLES

Cook in sugar without water whole unpeeled very fine apples until transparent. Then put the apples into jars filled with hot vinegar that has been boiled with honey, allspice and fresh rosemary. The jars should be hermetically closed, and the apples not served for a couple of months.

Cecil Beaton

ICED APPLES

(a Greek pudding, very Oriental)

Prepare a syrup with 2 cups sugar and 3/4 cup water and the rind of a lemon. Peel and cut in very thin slices 2 lbs. apples of a very good quality. Put them in the syrup and let them cook from 2 to 2 1/2 hours. Pour into a mould. Surround when removed from mould with a vanilla custard sauce. Decorate it with candied fruit. Serve very cold. Should be prepared the day before, or in the morning if served for dinner.

The Late Lord Berners

PUDDING LOUISE

Line a flan ring with short pastry, then put 6 layers of red currant jelly at the bottom, then the mixture of 3 1/2 ozs. sugar, 3 1/2 ozs. butter, 2 ozs. flour, and cook in a very moderate oven for 1 hour until very brown.

MARY OLIVER
London

WEDDING ANNIVERSARY ICE CREAM

Take 12 crystallised mint leaves, 1 cup crème-de-menthe, 1 oz. crystallised ginger, 1 quart thick cream and freeze.

FROM THE SAME SOURCE

BIRTHDAY ICE CREAM FOR ADULTS

Toast 2 slices of dark brown bread, spread *lavishly* with butter on both sides. Cut into small cubes. Cover with egg nog made of 2 eggs and 1 cup rum. Add 1 quart cream and freeze.

MADAME JOSEPH DELTEIL
Grabels

VERY GOOD CHOCOLATE MOUSSE

1/2 lb. sweet chocolate. 6 eggs.

Grate the chocolate and melt it in a frying pan with 3 tablespoons water over very low heat. Add the yolks of eggs previously stirred and mix well. Remove from heat and add the whites of eggs beaten stiff. Put into the serving bowl and into the refrigerator overnight. Always liked. Rather sponge-like.

PRINCESS D. DE ROHAN
London

CRÈME BRULÉE

Serves 4.

Stir, bring to boiling point, and boil for exactly 1 minute 2 cups heavy cream. Remove the cream from the fire. Pour it in a slow stream into 4 well-beaten egg yolks. Beat it constantly. Return the cream to the fire. Stir and cook it over a low flame until it is nearly boiling, or stir and cook it for 5 minutes in a double boiler. Place the cream in a buttered shallow baking dish and never stir again. Chill it well. Cover the cream with 1/4-inch layer of brown sugar. Place it under a broiler (keep the oven door open) to form a crust. Chill it again.

Mrs. Joseph A. Barry
New York and Paris

ORANGE AND LEMON DESSERT

Ingredients: 2 dozen lady fingers split, 1/4 lb. butter, 1 cup sugar, 1/2 pint heavy cream, 3 eggs beaten separately, juice of 1 large orange, juice of 1 lemon, 1 tablespoon orange rind, 1 teaspoon lemon rind, 1/4 cup chopped nutmeats ground very fine, rum.

Cream butter and sugar. Beat egg yolks with a whisk and add to creamed mixture. Add fruit juices, rind and nuts. Beat cream and add. Beat egg whites with a pinch of cream of tartar and add last.

Line a 1 1/2-quart porcelain *soufflé* mould with buttered waxed paper. Then line bottom and sides of mould with split lady fingers. Sprinkle with rum according to personal taste. Add half of mixture, then a layer of lady fingers, the second half of mixture and a ceiling of lady fingers with more rum. Put in freezing compartment of refrigerator for 2 or 2 1/2 hours (not more). Believe it or not, rich as this is, men guests often take two helpings. It is the rum that keeps it from being unmanly.

Brion Gysin

HASCHICH FUDGE

(which anyone could whip up on a rainy day)

This is the food of Paradise—of Baudelaire's Artificial Paradises: it might provide an entertaining refreshment for a Ladies' Bridge Club or a chapter meeting of the DAR. In Morocco it is thought to be good for warding off the common cold in damp winter weather and is, indeed, more effective if taken with large quantities of hot mint tea. Euphoria and brilliant storms of laughter; ecstatic reveries and extensions of one's personality on several simultaneous planes are to be complacently expected. Almost anything Saint Theresa did, you can do better if you can bear to be ravished by '*un évanouissement reveillé*.'

Take 1 teaspoon black peppercorns, 1 whole nutmeg, 4 average sticks of cinnamon, 1 teaspoon coriander. These should all be pulverised in a mortar. About a handful each of stoned dates, dried figs, shelled almonds and peanuts: chop these and mix them together. A bunch of *canibus sativa* can be pulverised. This along with the spices should be dusted over the mixed fruit and nuts, kneaded together. About a cup of sugar dissolved in a big pat of butter. Rolled into a cake and cut into pieces or made into balls about the size of a walnut, it should be eaten with care. Two pieces are quite sufficient.

Obtaining the *canibus* may present certain difficulties, but the variety known as *canibus sativa* grows as a common weed, often unrecognised, everywhere in Europe, Asia and parts of Africa; besides being cultivated as a crop for the manufacture of rope. In the Americas, while often discouraged, its cousin, called *canibus indica*, has been observed even in city window boxes. It should be picked and dried as soon as it has gone to seed and while the plant is still green.

BEVERAGES

REDVERS TAYLOR
Bishops Lydeard and London

SLOE GIN
(use unsweetened London gin, not Plymouth)

To each bottle of gin allow 1 pint sloes and 1/2 lb. rock candy (as white and clear as possible). Have ready two empty quart bottles. Prick the sloes with a fork (silver for preference). Put 1/2 pint sloes and 1/4 lb. rock candy, crushed fine, followed by 1/2 bottle gin into each bottle.

Allow to stand for three months, shaking every day. Then strain off through muslin and bottle. Seal the cork. Leave at least 1 year before drinking. The longer the better—at seven years it's a dream.

PRINCESS D. DE ROHAN
London

HOT TODDY FOR COLD NIGHT
(attributed to Flaubert)

2 jiggers Calvados.
1 jigger apricot brandy.
Warm over flame. Slowly pour in a jigger cream. Do not stir.
This is the recipe of the eighteenth-century *Auberge du Vieux Puits* at Pont Audemer.

FROM THE SAME SOURCE

DUBLIN COFFEE JAMES JOYCE

2 jiggers Irish whiskey in a balloon wine-glass.
1 teaspoon sugar.

Pour in black coffee, stir; as contents revolve, add jigger cream slowly in circular motion. Allow cream to float on top of coffee. Do not stir again.

Excellent for after-dinner conversation.

Miss Ela Hockaday
Dallas

EGG NOG OF THE COMMONWEALTH CLUB, RICHMOND, VIRGINIA

2 dozen eggs.
2 quarts cream.
1 quart whipping cream.
4 ozs. rum.
4 ozs. brandy.
1 1/2 lbs. sugar.
2 quarts whiskey.

Separate yolks and whites of eggs. In a large bowl beat thoroughly the yolks of the eggs, then add and mix well the sugar, adding slowly. Stir and heat mixture well and stir in well the whiskey. Add cream slowly and mix thoroughly. Beat whites of eggs till stiff and mix in thoroughly, then lastly the whipped cream.

This has been used in this club for more than a hundred years, I am told. Always served Christmas morning—and many other times!

PRESERVES AND A CHUTNEY

Redvers and Louise Taylor
Bishops Lydeard and London

ORANGE MARMALADE

Three oranges and 4 lemons. Cut into thin slices taking out the seeds. Put in 2 quarts water and let stand for 24 hours. Then boil for 1 hour and let stand again for 36 hours in a cool place. Add 4 lbs. sugar and boil for 1 hour or until it jellies.

From the Same Source

RHUBARB PRESERVE

Six lbs. rhubarb, 6 lbs. sugar and 6 large lemons. Cut the rhubarb in small pieces. Slice the lemons very thin. Put the fruit in a large bowl and cover with the sugar. Let stand until it has drawn out the juice. Then boil for about 3/4 hour. Do not stir more than necessary as its great beauty is in its not being all broken up. Place the leaf of a scented geranium in the bottom of each jar before bottling.

This Too from the Same Source

APPLE CHUTNEY

3 dozen cooking apples.
3 lbs. onions.
3 lbs. brown sugar.
1/2 lb. sultanas.
1/2 oz. chilis.
1/4 lb. salt.
2 quarts vinegar.
1 oz. ground ginger.
2 ozs. mustard seed.

Chop the apples and onions very small. Mix the whole together and simmer all day until it becomes a dark pulp.

XIII.

THE VEGETABLE GARDENS AT BILIGNIN

FOR FOURTEEN SUCCESSIVE YEARS THE GARDENS AT BILIGNIN WERE my joy, working in them during the summers and planning and dreaming of them during the winters. The summers frequently commenced early in April with the planting, and ended late in October with the last gathering of the winter vegetables. Bilignin surrounded by mountains and not far from the French Alps—from higher ground a few miles away Mont Blanc was frequently visible—made early planting uncertain. One year we lost the first planting of string beans, another year the green peas were caught by late frost. It took me several years to know the climate and quite as many more to know the weather. Experience is never at bargain price. Then too I obstinately refused to accept the lore of the farmers, judging it, with the prejudice of a townswoman, to be nothing but superstition. They told me never to transplant parsley, and not to plant it on Good Friday. We did it in California, was my weak reply. They said not to plant at the moment of the new or full moon. The seed would be as indifferent as I was, was my impatient answer to this. But it was not. Before the end of our tenancy of the lovely house and gardens at Bilignin, I had become not only weather-wise but a fairly successful gardener.

In the spring of 1929 we became tenants of what had been the manor of Bilignin. We were enchanted with everything. But after careful examination of the two large vegetable gardens—the lower on a level with the terrace garden in front of the house, and the other on a considerably higher level and a distance from the court and portals—it was to my horror that I discovered the state they were in. Nothing but potatoes had been planted the year before. Poking about with a heavy stick, there seemed to be some resistance in a corner followed by a rippling movement. The rubbish and weeds would have to be cleaned out at once. In six days the seven men we mobilised in the village had accomplished this. In the corner where I had poked, a snake's nest and several snakes had been found. But so were raspberries and strawberries.

A plan was made for plots and paths. The French are adroit in weeding and gathering from paths a few inches in width. It was difficult for me to accommodate myself to them. A list of what vegetables and when they were to be planted had to be made. We had brought with us sacks of seeds of all the vegetables Gertrude Stein and I cared for and some with which we would experiment. To do the heavy work a boy from the hamlet had been found. After fertilisers had been turned into the ground and the topsoil raked to a powder, with a prayer the planting commenced. The seeds for early gathering had scarcely been disposed of when it was time to plant the slips bought from farmers' wives in the square at Belley at the Saturday-morning market. There were two horticulturists in Belley, one an over-educated ambitious pretentious youngish man who never ceased believing he was to be the next Minister of Agriculture, and an old woman who had no more preparation for her work than experience which had taught her little. The slips we had from them were not all vigorous; we planted twice as many as we intended to grow.

The wind blew the seeds from the weeds in the uncultivated fields that surrounded the gardens. The topsoil was of made earth and heavy rains commenced to wash it away. These were the two disadvantages we should have to overcome. The weeds remained a tormenting, backbreaking experience all the summers we spent at Bilignin. After the autumn gathering was over, the topsoil would be renewed. Fortunately there was plenty of water. Only once was there a drought, when the ox carts brought water in barrels from the stream in the valley below. For watering we had bought three hundred feet of hose.

The work in the vegetables—Gertrude Stein was undertaking for the moment the care of the flowers and box hedges—was a full-time job and more. Later it became a joke, Gertrude Stein asking me what I saw when I closed my eyes, and I answered, Weeds. That, she said, was not the answer, and so weeds were changed to strawberries. The small strawberries, called by the French wood strawberries, are not wild but cultivated. It took me an hour to gather a small basket for Gertrude Stein's breakfast, and later when there was a plantation of them in the upper garden our young guests were told that if they cared to eat them they should do the picking themselves.

The first gathering of the garden in May of salads, radishes and herbs made me feel like a mother about her baby—how could anything so beautiful be mine. And this emotion of wonder filled me for each vegetable as it was gathered every year. There is nothing that is comparable to it, as satisfactory or as thrilling, as gathering the vegetables one has grown.

Later when vegetables were ready to be picked it never occurred to us to question what way to cook them. Naturally the simplest, just to steam or boil them and serve them with the excellent country butter or cream that we had from a farmer almost within calling distance. Later still, when we had guests and the vegetables had lost the aura of a new-born miracle, sauces added variety.

In the beginning it was the habit to pick all vegetables very young except beetroots, potatoes and large squash and pumpkins because of one's eagerness, and later because of their delicate flavour when cooked. That prevented serving sauces with some vegetables—green peas, string beans (indeed all peas and beans) and lettuces. There were exceptions, and for French guests this was one of them.

GREEN PEAS À LA FRANÇAISE

Put in a saucepan over medium heat 4 cups shelled green peas, 12 spring onions, a bouquet of 1 sprig of parsley and several stalks of basil, 1/4 cup butter, 1/4 teaspoon salt, 1/2 teaspoon sugar, 1 white lettuce cut in ribbons and 4 tablespoons water. Bring to a boil, cover and reduce heat gradually to low. Before serving remove bouquet and add 4 tablespoons butter. Tip saucepan in all directions to melt butter. Serve at once.

Or this one:

GREEN PEAS À LA GOOD WIFE

Put 12 young onions in a saucepan over medium heat with 3 tablespoons butter and 1/2 cup fat back of pork previously boiled for 5 minutes, drained and cut in cubes of 1/2 inch. When the onions are lightly browned, remove with the cubes of back fat. In the saucepan stir 1 tablespoon flour. Mix well and gradually add 1 3/4 cups veal *bouillon*. Allow to boil for 15 minutes and salt. Then add 4 cups shelled green peas, the diced pork fat and the onions. Cover and cook from 15 to 25 minutes according to the size and age of the peas.

Here is still another recipe for

GREEN PEAS AS COOKED IN PROVENCE BY THE FARMERS

Put in an earthenware casserole, which has a well-fitted cover and which is fireproof, 5 tablespoons olive oil and 1 medium-sized onion. Brown the onion over low heat. Then add 6 medium-sized potatoes cut in 1/3-inch slices. Stir the potatoes until they are covered with oil, and 4 cups boiling water, 4 cups shelled green peas, 3 cloves of crushed garlic, a bouquet of half a laurel leaf, a twig of thyme and a slice of fennel or a bunch of fennel greens, 1/2 teaspoon salt, 1/4 teaspoon pepper and 1/2 teaspoon saffron. Cover and boil gently. When the peas are tender slide on the surface of the juice 4 eggs. They must not touch each other. Gently pour 2 tablespoons of the juice over the yolks of the eggs to form a film. Cut 2 slices of French bread 1/3 inch thick and place on the plates on which the peas and the sauce are to be served.

In summer there were about thirty different vegetables which had to be examined and gathered every morning as well as the berries. There were more berries than we could eat either in desserts or fresh, so I made jam of the little wood strawberries and jelly of the raspberries. The French have a perfect way of making

STRAWBERRY JAM (2)

Put equal weight of sugar and berries into an earthenware bowl. Stir gently not to bruise the berries until they are coated with sugar. Put aside for 24 hours. Pour off sugar and juice that has exuded from berries. Place over medium heat in enamelled saucepan or pot and bring to a boil. Boil gently for 3 or 4 minutes. Remove from heat. Skim and place berries in glasses. Replace syrup over heat and boil gently until spoon is thickly coated. Fill glasses.

Ordinary strawberries may be cooked in the same way. They must of course be hulled or stripped and preferably not washed. They will require from 12 to 15 minutes boiling, depending upon their size and their ripeness. The hulls of the little berries remain on the stalks when the berries are picked.

Most of our men guests had their breakfasts served on the terrace. The breakfast trays were my pride, though the linen and porcelain were simple. In the market place in Chambéry we unearthed some amusing coloured glassware, 1840–1850, from Savoy, not at an antique dealer's but in a store that sold glassware. We bought all there was. Berries, fruits, salads and vegetables served in them were subjects for still-life pictures. For French guests an abbreviated American breakfast was served, somewhat more ample however than their usual coffee and rolls or *croissants* with butter, jam, marmalade or jelly. *Croissants* are a delicious accompaniment to breakfast or to tea.

CROISSANTS OR CRESCENTS

Heat 1/2 cup milk. When it is warm mix into it 1 package of compressed yeast. Sift 1 cup flour and mix with the yeast to make a sponge. Allow to rise, for about 1/2 hour. Sift 3 cups flour into a large bowl. Put the yeast at the bottom in the centre of the bowl and gradually work in 3 1/2 cups milk and the flour. Put aside until it has risen to twice its size. Then place it on a lightly floured board and knead thoroughly until the dough no longer sticks to the hands. Roll out and place 1/4 cup butter, which has been worked with the hands into a square, in the centre, fold the dough from the sides to meet in the centre. Roll with the hands into a ball and keep in a cool place for several hours or even for the night. Then roll out again and divide into pieces the size of an egg. Roll each one into a cylinder and put aside for 10 minutes. Then very lightly roll them out to 1/3 inch thickness. Roll from one corner, bend into the shape of crescents and put aside for 35 minutes. Place on a lightly buttered baking sheet, paint with pastry brush dipped in slightly beaten egg mixed with 1 tablespoon water. Bake in preheated 425° oven.

These are a typically French bread, though they were created in Austria.

String beans with a sauce is a desecration, especially string beans grown in France. Even in my well-loved California they were less tender, less flavoursome, and not as free from strings. Still they could be cooked—with our French friends in mind—disguised to my mind as

STRING BEANS IN THE PROVENÇAL MANNER

Heat 3 tablespoons olive oil in frying pan. Add 1/4 cup capers, 1/4 cup boned anchovies, 1 crushed clove of garlic and 4 cups string beans previously boiled until tender in boiling water with 3/4 teaspoon salt and 1/4 teaspoon pepper in an uncovered saucepan. Toss them in the frying pan until they are well mixed with the various ingredients in it. Serve hot with chopped parsley and fine chopped spring onion sprinkled on them.

And this way:

STRING BEANS BROWNED IN THE OVEN

Boil in salted water uncovered 4 cups string beans. Do not overcook, but when tender remove from heat and drain thoroughly. Butter a fireproof earthenware dish. Sprinkle generously with grated Swiss cheese. Mix 1 1/2 cups *Béchamel* sauce with 2 yolks of eggs. Pour 1/4 cup of this on the cheese in the earthenware dish, then the string beans and on them the rest of the sauce. Sprinkle 1/3 cup grated Swiss cheese on the string beans, dot with 3 tablespoons melted butter and put in preheated 450° oven for 15 minutes.

These are very nice ways to cook string beans but they interfere with the poor vegetable's leading a life of its own. As for the cooking of lettuces, with or without sauce, the easiest way to accept this is to consider lettuce as two vegetables, raw lettuce and cooked lettuce. To me cooking lettuces is the sacrifice of the innocents. If cooked they must be, this way is always received with appreciation:

LETTUCE IN RIBBONS WITH CREAM

For 4 cups shredded lettuce put 4 tablespoons butter in a saucepan over medium flame, put the shredded lettuce in the butter and turn with a wooden spoon until all the lettuce is covered with butter. Reduce the heat to low, add 1/4 teaspoon salt. Cover and simmer until all the moisture in the lettuce has been absorbed. Add 1 1/2 cups thick cream sauce and 1 teaspoon onion juice. Stir well together. Place on a serving dish in a mound, and surround by large triangular *croûtons*.

BRAISED LETTUCE WITH ITALIAN SAUCE

Wash and dry 4 large lettuces having removed the green outer leaves. Put into 4 tablespoons butter melted in a saucepan. Stir, reduce heat and cover. Simmer until all the water has evaporated, then add this sauce.

Chop coarsely 1/3 cup mushrooms. Place mushrooms in saucepan over low heat with 2 tablespoons olive oil and 2 tablespoons butter, 1 tablespoon chopped onion and 1/2 tablespoon chopped shallot. Mix well and add 1/3 cup hot dry white wine. Cover and simmer for 10 minutes, add 1/4 cup tomato *purée*, 1/4 teaspoon salt. Add 1/4 cup butter in small pieces. Melt and mix by tipping the saucepan in all directions. Do not allow to boil.

This is a change—indeed a violent one.

There were lettuces for early gathering and for summer. For autumn not only Batavia and Romaine, but an endless variety of vegetables to be served raw or cooked. For the early raw vegetables we waited for tomatoes to ripen. Neither Gertrude Stein nor I cared for raw root vegetables, nor did we care for spinach in salad. Spinach and sorrel—blessed perennials that they are—could be gathered early. The second or third year at Bilignin I dug up the row of sorrel keeping only one plant. Sorrel is not a delectable dish, but from time to time a few leaves in a mixed salad add an agreeable tang, or chopped very fine and sprinkled on cold fish they have their place. At Bilignin a friend looking over the two vegetable gardens was surprised that there was no Henry IV spinach growing in either of them. It was the first time it had been spoken of to me. So obediently the next year Henry IV spinach was planted. There was no difference between it and ordinary spinach except that the Royal kind, as we got to calling it, grew on poles and one could gather it without stooping. On the other hand it bore less prolifically. For several years it was planted as a curiosity and amusement for our American friends. After three or four years it was no longer planted.

There was still another spinach which grew in the hot summer months without going to seed, called in French *Tetredragon*, which was known in the only American seed catalogues in which it was listed as New Zealand spinach.

The good Madame Roux, who in these early days of my apprenticeship would come into the garden when I was at work, interrupted the washing and ironing she was doing to give me tactful advice. From her I learned all that was taught me. It was not until the Occupation that she had the satisfaction of seeing the fruits of her effort. *À propos* of the New Zealand spinach, she said it required a good deal of fertiliser, that what suited it best was *la crème de la crème*, and when she saw that I had no idea what that was, she explained that it meant pig's manure. She came into the garden with a large wheelbarrow of it next morning, and having secured a second one, we hastily found space for two lines. The leaf is small and thick, shaped like the leaf of the ivy, and the plant has a tendency to creep and to climb. In the Bugey it grows on short poles, but it is wayward. Deliciously tender when cooked, it is best when boiled as it is, without the addition of water after washing. Ten minutes in a covered saucepan and drained, put under the cold-water tap and drained again, the water thoroughly pressed out and returned to the saucepan—that is the right way to cook it. Then salt should be added and sour cream or butter in quantity. A half teaspoon of salt, and a pinch of ground nutmeg or a pinch of ground ginger or both, give the necessary flavour. There are so few people who care for spinach, but this way of preparing it has seduced a number of my friends.

From the United States every year a kind friend sent a little packet of sweet-corn seed grown and gathered by his mother. It was a great treat for us. At that time there was no table corn in France. The French grew corn for animals—in the Bugey, for the chickens. When it was known that we were growing it and eating it, they considered us savages. No one was seduced by the young ears we gave them to taste. But what did astonish and please them were the giant globe tomatoes, not only red but yellow and white, and the very large Chinese globe egg plants, which held nearly a pint of stuffing each.

It was after seeing gumbo (or okra) growing in Méraude Guevara's garden in the south of France that gumbo seeds came to Bilignin. The plants flourished, almost alarmingly so. We couldn't eat half of what I gathered, and it was quite beyond our budget to keep the number of chickens with which to cook them. There were few lobsters and crabs in a small town so far from the sea as Belley. They were used, then, with river and lake fish, with veal, and in a mixed vegetable stew—not because the vegetable stew requires more than a dozen different ones, but because one has to commence to eat them in rotation the next day.

This delicious dish does not include gumbos, but 3/4 cup of them thinly sliced and 1 cup hot *bouillon* could be added.

MACÉDOINE OF VEGETABLES

Cooked with freshly gathered spring vegetables this mixed-vegetable dish is exquisite.

In a fireproof pot which has a well-fitting cover put 3 tablespoons butter and 3 medium-sized onions cut in rings. When brown, add 6 hearts of lettuce, 2 cups shelled sweet peas, 2 cups string beans, 2 small round carrots, 6 round turnips, 1 1/2 cups small new potatoes, 2 cups asparagus tips and 1 cup butter. Cover and cook over low flame. Shake the pot frequently and stir gently from the bottom occasionally with a wooden spatula or spoon. Be careful that the vegetables do not burn or scorch. After 1 3/4 hours add salt and pepper and cook for 1/4 hour—2 hours in all.

This is one of the very best of vegetable dishes.

This is a richer version of it:

MACÉDOINE OF CREAMED VEGETABLES

Boil 1 1/4 cups shredded carrots, 1 3/4 cups shredded turnips and 1 3/4 cups shelled peas in 1 quart milk. Boil 1 3/4 cups string beans and 1 3/4 cups asparagus tips in salted water. Steam 3 large potatoes, peel and dice. Do not overcook any of these vegetables. In a saucepan over low heat melt 2 tablespoons butter, add 1 tablespoon flour, stir with wooden spoon. Slowly add the hot milk that remains after cooking the carrots, turnips and peas. Stir carefully so that the sauce is perfectly smooth. Allow to simmer for 10 minutes, stirring constantly, and then add 1 cup heavy cream. Bring to the boil and mix with vegetables, and salt and pepper. Bring to the boil again, stir to mix but do not allow to boil. Remove from flame and add 2 tablespoons butter. Do not stir but tip saucepan in all directions to mix. Serve at once.

Also an exquisite dish.

The summer vegetables were often cooked simply, but in late August and through the autumn more elaborately. Spiced and richer ways seemed appropriate. Often fresh herbs were boiled with them, or chopped, for those cooked in the oven. My taste for garlic is pronounced—friends say exaggerated—and there never seems to be too great a flavour of basil in vegetables, fish or meat. This is for country cooking; cooking in town does not admit of so much condimenting, or spicing for that matter.

Italian squash (zucchini) and egg plant (aubergine) can be prepared in the same way as

STUFFED BRAISED PEPPERS

Put through the meat chopper 1 cup raw lean veal and 1 cup ham. Mix well with the yolks of 2 eggs, 2 tablespoons fresh basil chopped fine, 2 tablespoons white wine, 1 tablespoon parsley chopped fine, salt and pepper, 1 cup chopped mushrooms previously cooked in 2 tablespoons butter, the juice of 1/2 lemon, salt and pepper, 3 tablespoons grated Parmesan cheese, 1/2 teaspoon cumin powder and 1/4 teaspoon saffron. Cut in half, lengthwise, 3 red and 3 green sweet peppers, remove seeds and boil for exactly 2 minutes. Drain and wipe dry. Fill the peppers with the stuffing. Melt 5 tablespoons butter in a casserole with a tight-fitting cover. Place the stuffed peppers, not overlapping, in it. Moisten with a cup of hot chicken or veal stock. Add 2 cups chopped tomatoes. Cook covered over medium heat for 20 minutes. Remove peppers, place alternating yellow and red peppers on serving dish. Reduce sauce for a few minutes, then strain over peppers and serve.

This is a pleasant change from the too-frequently-served vegetable browned in the oven. The flavours marry well.

We were very proud of the artichokes the garden produced, too proud to do much to them until late autumn, when we ate this version of

ARTICHOKES À LA BARIGOULE

For 4 large artichokes put in a casserole with a tight-fitting cover over medium heat 5 tablespoons olive oil, 1 chopped onion and 2 diced carrots. Place the artichokes on these, leaves up, salt and pepper. Pour a little of the oil over the artichokes. Cover, stirring occasionally. When the carrots and onions begin to brown add 1 cup white wine, 2 cloves of garlic and 2 twigs of fresh rosemary or 1 teaspoon powdered rosemary. Cook over low flame until a leaf can be removed from artichokes. Serve hot with sauce poured over artichokes.

Oyster plants (or salsify), beetroot and Jerusalem artichokes can be made appetising, though not delicate. They were planted each year, though one wondered why. The Jerusalem artichokes are redeemed by their agreeable substance. Oyster plants have no excuse for existing: they are long, too clean, have little flavour and are deservedly unpopular. Here the sauce justifies the time spent in preparing

OYSTER PLANT (SALSIFY) WITH WINE SAUCE

Melt 4 tablespoons butter in a casserole with a tight-fitting cover. Lightly brown in it 1 cup diced raw veal, 1 cup diced ham, 1 diced medium-sized carrot, 1/2 finely chopped medium-sized onion, 1 tablespoon finely chopped parsley, 1 of tarragon and 1 of chervil. Add 1/2 cup hot dry white wine and 1 cup hot *bouillon.* Bring to a boil. Add a pinch of powdered laurel and 1/4 teaspoon pepper. Cover and cook gently for 2 hours. Clean and boil the oyster plant in salted water until nearly tender. Drain and tie in bunches like asparagus. Put in casserole and simmer for 1/2 hour. Add the juice of 1 lemon. Remove from flame and add 3 tablespoons butter in small pieces. Do not stir but tip casserole in all directions until butter is melted. Untie the oyster plants. Arrange neatly on preheated serving dish, pour the sauce over the oyster plants, and serve at once.

Celery root (celeriac) can be cooked in the same way. They will require not more than 1/4 hour to boil and should, after being well drained, be cut in strips like potatoes for frying.

As for beetroots, their excuse for being is the fine colour they add to pale dishes. As a vegetable this recipe combines the decorative with the tasty.

PURÉE OF BEETROOT

Bake beetroots in oven till quite soft, peel and mash through strainer with a potato masher, add one-third their volume of thick cream sauce. Place over low heat in a casserole, add salt and pepper. When about to boil, add 1 tablespoon butter cut in small pieces to 2 cups *purée.* Do not allow to boil, do not stir but tip casserole in all directions. Serve in a mound on preheated dish. Sprinkle with chopped parsley. Or serve as a border around veal or pork roast.

Besides all the vegetables in the two gardens, there were the fruits and berries. Besides all the strawberries there were the raspberries which bore from the end of May all through the summer until December—the first snow did not discourage them. They were in a protected corner and were a lovely sight, growing as grape vines do in Lombardy. Attached to three rows of wire, each root was allowed six branches, three on either side. When one came upon them unexpectedly, one did not know what all the pendent clusters of colour could be. They never seemed real to me, but a new and joyous surprise each morning. Every care they needed—and little enough did one pay for their beauty and for their incredible fertility—was more reward than labour. In spring the branches were tied on to the wires, and again later, so that the weight of the berries would not break the branches. Later still, there were new shoots from the root, and three or four of the healthiest were kept for the next year's bearing. The plants that were there when we came produced red berries. After we had secured a lease, we planted forty-eight white raspberries in the upper garden. They didn't thrive as well as the red, perhaps. As it was farther away from water and it was an exhausting effort to drag the hose so far, the upper garden had been chosen for the commoner vegetables—potatoes, pole beans and the marrow, squash and cucumber family. There was not room for more raspberries in the lower garden. The white ones were planted in lines near the currants, red, white and black, and the gooseberries. Unquestionably they felt they were not favoured, nor were they. They were not in the sun all day and never received the same attention as the red family below. The berries were much smaller but much sweeter. What a happy life it would be only to cultivate raspberries.

In France as much attention is given to black currants as to red, and there are always a few bushes of white ones in every proper garden. The bushes of black currants have a very strong agreeable fragrance. Of them can be made a pleasant summer drink.

BLACK-CURRANT LIQUEUR

Wash and drain 1/2 lb. raspberries and 3 lbs. black currants. Mash them thoroughly. Cover with a cloth and put aside in a cool place for 24 hours. The next day add a handful of black-currant leaves that have been washed and dried, and 1 quart 90 per cent alcohol. Cover the bowl with a plate and put aside for 24 hours. On the third day place a sieve with a piece of fine linen over a bowl. Pour the fruit and the alcohol through, mashing with a heavy pestle or a potato masher. Put 3 lbs. sugar and 3/4 quart water in an enamelled saucepan. Stir until it commences to boil, over low heat. Boil for 5 minutes. Put aside until completely cold. Then add the syrup to the juice of the berries and the alcohol. Allow this to stand for several hours. Fit a filtering paper into a funnel and pour through it to fill the bottles. A filtering paper is bought at a good chemist's. This makes about 2 3/4 quarts.

This is not only served as a liqueur. In the Bugey it was poured into a glass filled with shaved ice. It was a very refreshing summer drink. A *mocha* (coffee) liqueur was also used for an iced drink. There were vulgarians who put whipped cream on top, and it was then called *à l'Américaine*—disparagingly, one is forced to acknowledge.

The gooseberries in France are four or five times larger than those grown in the United States, and very much sweeter. We grew a species that were raspberry coloured. They are cultivated like olive trees, the centre growths are removed as they appear—to give light and sun. Every year hornets would make a nest in the trunk of one of the bushes and with a sharp knife I would have to cut it out. Wasps, hornets and bees rarely sting me, though my work with them has always been aggressive. Gertrude Stein did not care for any of them, nor for spiders, centipedes and bats. She had no violent feeling about them out of doors, but in the house she would call for aid. The instruments for getting rid of them were determination, newspapers, a broom and pincers. These were always effective.

A charming story of wifely and husbandly devotion was that of two of our friends. She did not wish her husband to be bored, annoyed or worried. When they were first married she allowed him to believe that she was very much afraid of spiders. Whenever she saw him disturbed she would call him with a wail, Darling, a spider: there, darling—don't you see it. He would come flying with a handkerchief, put it on the spot indicated, and, gathering up the imaginary spider, would throw it into the garden. The wife would uncover her face and with a sigh say, How good and patient you are, dearest.

Rhubarb grew in the upper garden too. Two or three spring rhubarb tarts each year were not worth the space the rhubarb occupied, so they were rooted up and put on the compost heap. Melons too were not attempted after the first year. They required too much care, putting under glass and suppressing trailers and buds. In Paris we had a small room Gertrude Stein called the Salon des Refusés after the famous one at which the Impressionist painters showed their pictures the year when they had been refused admittance at the regular *salon.* Ours held the pictures Gertrude Stein refused—that is, pictures she had bought to find out what she felt about them and stored there when she found they did not interest her. In the garden it was simpler. When the *refusés* were rooted up and put on the compost heap, it caused no feeling from anyone involved.

Like all beans, pole beans flourish in French soil. In spite of reducing the number planted, there were always too many to gather. Then too they grew too high, nothing stopped them. Finally they were clipped at the top. To gather them was a problem not only because of the time it took but because of the narrow paths between rows. It wasn't practical to leave some steps out of doors, but by carefully balancing three very strong narrow wine cases one on top of the other and overlapping, the difficulty seemed overcome until one day we all came down in a heap dragging a pole and its garniture with us! A bruised leg put an end for a while to the beanstalk problem.

Fruits were neither as plentiful nor of so good a quality as the berries. When we came to Bilignin there was one pear, one plum and one apple tree in the garden, and all of them were old. With the landlord's permission we had the plum tree removed at once. Everything was done, but to no avail, to save the deep-rose-coloured pear tree. That left one excellent large apple tree. For an orchard it was not excessive—Gertrude Stein spoke of it as The Nucleus. We waited for three years until we secured a lease to plant apples, peaches, apricots and nectarines. The French like to plant fruit trees in the old-fashioned way, on the sunny side of the wall. There were only two such walls free, and a fine old laurel tree covered a part of one of them.

The laurel was a constant delight. There was a nameless mauve rose that cried for a border of laurel leaves. A bouquet of them was always in the bedroom of our young guests, writers and painters and occasionally musicians, as a symbol of a future wreath. None of them remarked the leaves.

The peaches, apricots and nectarines were not for long—they deteriorated in three or four years. It was not a soil or a climate that suited them. We were slow to learn this. As the apples were thriving, more of them replaced the fruits we were discarding. The famous Calville, of which the equally famous Calvados, an apple brandy, is not made, and the resembling Belles-Fleur Jaune, grew quickly and well. But fruit trees on a wall are not prolific. The Calvilles, sold by the piece in Paris, grow no more than forty to the tree. It was necessary to find a commoner, more hardy apple and to plant it in open ground—what the French, looking facts in the face, call open to the winds. And then we finally had an apple orchard. The forcing beds were returned to their sheltered home against the wall.

When autumn came, the last harvest was so occupying that one forgot that it meant leaving the garden for the return to Paris. Not only did the winter vegetables have to be gathered and placed to dry for a day before packing, but their roots and leaves had to be put on the compost heap with manure and leaves and packed down for the winter. The day the huge baskets were packed was my proudest in all the year. The cold sun would shine on the orange-coloured carrots, the green, yellow and white pumpkins and squash, the purple egg plants and a few last red tomatoes. They made for me more poignant colour than any post-Impressionist picture. Merely to look at them made all the rest of the year's pleasure insignificant. Gertrude Stein took a more practical attitude. She came out into the denuded wet cold garden and, looking at the number of baskets and crates, asked if they were all being sent to Paris, that if they were the *expressage* would ruin us. She thought that there were enough vegetables for an institution and reminded me that our household consisted of three people. There was no question that, looking at that harvest as an economic question, it was disastrous, but from the point of view of the satisfaction which work and aesthetic confer, it was sublime.

Our final, definite leaving of the gardens came one cold winter day, all too appropriate to our feelings and the state of the world. A sudden moment of sunshine peopled the gardens with all the friends and others who had passed through them. Ah, there would be another garden, the same friends, possibly, or no, probably new ones, and there would be other stories to tell and to hear. And so we left Bilignin, never to return.

And now it amuses me to remember that the only confidence I ever gave was given twice, in the upper garden, to two friends. The first one gaily responded, How very amusing. The other asked with no little alarm, But, Alice, have you ever tried to write. As if a cookbook had anything to do with writing.

INDEX OF RECIPES